THE GUIDE TO
ENTREPRENEURSHIP

Michael Szycher, PhD

THE GUIDE TO
ENTREPRENEURSHIP

CRC Press
Taylor & Francis Group
Boca Raton London New York

CRC Press is an imprint of the
Taylor & Francis Group, an **informa** business

CRC Press
Taylor & Francis Group
6000 Broken Sound Parkway NW, Suite 300
Boca Raton, FL 33487-2742

Printed on acid-free paper
Version Date: 20140826

International Standard Book Number-13: 978-1-4822-0907-5 (Paperback)

Library of Congress Cataloging-in-Publication Data

Szycher, M. (Michael)
  The guide to entrepreneurship : how to create wealth for your company and stakeholders / Michael Szycher, Ph.D.
    pages cm
  Summary: "Whether you work for an established company and want to trailblaze new products (intrapreneurship), or want to establish a new venture (entrepreneurship), this guide will serve as an invaluable resource. The book explores what constitutes entrepreneurial timber and the leadership skills to raise all the needed capital. Most new venture business failures arise from two sources: (1) opportunity is not real, or (2) the means required to pursue the idea is not available. This book will provide the aspiring entrepreneur with a "how-to-guide" to success, with concrete action plans"-- Provided by publisher.
  Includes bibliographical references and index.
  ISBN 978-1-4822-0907-5 (paperback)
  1. Entrepreneurship. 2. New business enterprises. 3. New products. 4. Success in business. I. Title.

HB615.S99 2014
658.4'21--dc23
2014004838

Visit the Taylor & Francis Web site at
http://www.taylorandfrancis.com

and the CRC Press Web site at
http://www.crcpress.com

This book is dedicated to my wife, Laurie, whose unwavering support was my inspiration to complete this *Guide*, and to my granddaughter, Arielle, and my grandson, Jason. May they both be inspired by the contents of this *Guide* as they grow up.

Contents

Acknowledgments xvii

Introduction xix

1   **Entrepreneurial Timber** 1
    1.1   Introduction 1
    1.2   Entrepreneurial Anatomy 3
    1.3   The Entrepreneurial World 3
    1.4   Entrepreneurial FAQ 5
    1.5   Success Characteristics 5
    1.6   Failure Characteristics 5
    1.7   Risk Is a Four-Letter Word 11
    1.8   Challenges of Entrepreneurship 11
    1.9   The Ten Commandments of Entrepreneurship 12
    References 15

2   **Intrapreneurship: Corporate Entrepreneurship** 17
    2.1   Introduction 17
    2.2   Definitions 17
          2.2.1   Kelly Johnson's 14 Rules of Skunk Works 18
    2.3   Theory of Intrapreneurship 20
          2.3.1   The 3M Illustration 21
    2.4   Characteristics of Intrapreneurs 24
    2.5   External and Internal Business Environment 25
    2.6   Culture regarding Innovation 26
    2.7   Corporate Support for Internal Business Creation 27
    2.8   Critical Issues in Intrapreneurship 28

2.9 The Spin Zone 30
    2.9.1 The Spin-Out 31
    2.9.2 The Spin-Off Reorganization 33
    2.9.3 The Equity Carve-Out 34
2.10 Intrapreneurship in Academia 37
2.11 The Ten Commandments of Intrapreneurship..............38
References 39

3 Women and Entrepreneurship 41
3.1 Introduction 41
3.2 Global Impact of Women Entrepreneurs.........................41
3.3 Personal Challenges Faced by Female Entrepreneurs....43
3.4 Women and Success...43
    3.4.1 Defining Success..44
    3.4.2 Twenty-Five Leading Female CEOs 44
3.5 Hurdles to Overcome 45
3.6 Role of Angel Groups..46
    3.6.1 Angel Capital Market and Female Entrepreneurs...47
3.7 The Green Alliance. A Female-Owned Business.............47
    3.7.1 Profile of Sarah Brown, Owner, Director 47
    3.7.2 The Green Alliance Business Model.......................48
    3.7.3 Business Conceptualization.....................................49
    3.7.4 Green Alliance Mission 50
    3.7.5 The "Green" Business Enterprise 50
3.8 An Interview with Sarah Brown 51
    3.8.1 The Importance of Emotional Support...................52
    3.8.2 Don't Undersell Yourself 53
    3.8.3 Overcoming Perceptions, Gaining Respect, and Finding
         Success 54
    3.8.4 Putting It in Perspective 54
References 54

4 The Entrepreneurial Environment.....................................57
4.1 Introduction 57
4.2 Leadership Qualities...59
    4.2.1 Leaders vs. Managers 59
    4.2.2 The Centrality of Leadership in a StartUp............61
    4.2.3 Leadership Is Situational 62

4.3 Establishing Your Founder Team 64
    4.3.1 Elements of Skilled Teams 64
    4.3.2 Establishing Your Board of Directors 65
    4.3.3 Selecting Your Board of Advisors 66
4.4 Authority, Power, and Influence 68
4.5 The Founder and Organizational Politics 69
    4.5.1 Power, Politics and Influence 71
    4.5.2 Organizational Persona 72
4.6 The Dominant Coalition 74
    4.6.1 Dominant Coalition as Change Agent 75
        4.6.1.1 Theory of Change 76
        4.6.1.2 Force Field Analysis 77
4.7 Leaders Need Followers 77
    4.7.1 Current Views of Followership 79
    4.7.2 The Followership Universe 80
4.8 Time Management 80
    4.8.1 The Urgent-Important Matrix 81
4.9 Conflict Management 82
4.10 Entrepreneurial Rewards 84
4.11 Entrepreneurial Types 85
4.12 The Ten Commandments of Leadership 86
References 87

5 New Venture Creation 89
5.1 Introduction 89
5.2 Corporate Life Cycle 90
5.3 Global Fortune 500 Annual Turnover 91
5.4 Corporate Life Cycles 91
    5.4.1 Greiner's Model of Organizational Growth 93
    5.4.2 Churchill and Lewis Growth Model 94
    5.4.3 New Product Life Cycles 96
5.5 A Small Company Is Not a Little Big Company 99
    5.5.1 Cash Is King 101
5.6 Your Innovative Organization 103
    5.6.1 Types of Innovation 103
5.7 Creating Your Own Innovation Culture 106
5.8 Strategic Planning for StartUps 107
5.9 Your Value Chain Analysis 108
5.10 Your Value Proposition 109

5.11 SWOT Analysis 112
5.12 The Early Adopters... 113
5.13 Bridging the Valley of Death 115
References 115

6 Financing Your Dream... 117
6.1   Naming Your Baby.. 117
      6.1.1 Bad Company Names... 117
      6.1.2 Bad Product Names... 118
6.2   Incorporating Your New Venture 118
      6.2.1 Characteristics of Corporations 119
6.3   Selecting the Optimal Legal Structure 119
6.4   Legal Powers of Corporate Executives 121
6.5   Creating Value for Your Stakeholders.................................. 123
6.6   Employee Benefits.. 123
6.7   The 3 Fs, Angels, and VC.. 124
      6.7.1 Friends and Family.. 125
      6.7.2 Angel Investors 126
      6.7.3 Venture Capital 126
6.8   Principles of Raising Capital... 127
6.9   Persuasive Business Presentations 128
      6.9.1 Rookie Mistakes.. 131
6.10 Your Elevator Pitch 131
      6.10.1 Must Haves 132
      6.10.2 Brief Descriptions... 133
      6.10.3 Last Three Bits of Advice 133
6.11 Estimating StartUp Costs.. 133
6.12 Valuing You and Your Team... 133
6.13 Valuing Your New Venture (Calculating Pre-Revenue
      Valuation).. 134
      6.13.1 Basic Calculations... 134
      6.13.2 Valuation Examples 135
References 136

7 Organizational Structure 137
7.1   Introduction... 137
7.2   The Founding Team.. 137
7.3   Hiring Your Executive Team.. 139
      7.3.1 Teamwork Not Titles 140

7.3.2 Team Discipline 140

7.3.3 Focus on Outcomes, Not Activities 141

7.4 Too Many Chiefs, and Not Enough Productive Indians 141

7.5 Accountants and Lawyers 142

7.5.1 Accountants 142

7.5.2 Lawyers 142

7.6 Managing Managers 143

7.6.1 How Is Managing Managers Different from
Managing Projects? 143

7.6.2 A Roadmap for Managing Managers Effectively 143

7.7 Selling Your Vision to Your Team 145

7.8 Fail as Fast as You Can 146

7.9 Overcoming Organizational Inertia and Defenses 147

7.9.1 Organizational Inertia 148

7.9.2 Organizational Defenses 149

7.10 An Old Aphorism 150

References 151

8 **Power Negotiations** **153**

8.1 Business Negotiations 101 153

8.1.1 Negotiation Wheel 154

8.1.2 Negotiation Strategies 154

8.1.3 Distributive Negotiation 156

8.1.4 Integrative Negotiation 157

8.2 The Entrepreneur as Chief Negotiator 157

8.2.1 StartUp Negotiator Fears 158

8.2.2 The Neophobia Syndrome 158

8.2.3 If You Expect More, You Get More 159

8.3 Knowing Your Negotiation "Counterpart" 159

8.4 Your First Negotiation 160

8.4.1 Aligning Interests 161

8.5 Planning Your Negotiation Success 161

8.6 The Negotiation Agenda 162

8.6.1 Preparing the First Draft of Any Document 163

8.6.2 Measuring Success 164

8.6.3 Negotiation Strategies 164

8.6.4 Negotiation Styles 166

8.6.5 Negotiation Tactics 166

8.7 Laws of Power Negotiation 169
    8.7.1 The Chicken or Egg Cycle 170
    8.7.2 The Negotiating Table 171
    8.7.3 How Is Your Meeting Progressing? 172
    8.7.4 Power Negotiator's TIR 173
8.8 Managing Negotiation Conflict 173
8.9 Secrets of Power Presentations 174
8.10 Secrets of Closing the Deal 175
8.11 The Ten Amendments of Negotiations 177
References 178

9 Marketing and Sales 181
9.1 Introduction 181
9.2 Selling Your Innovation 183
    9.2.1 Needs, Wants, and Demands 184
9.3 Analyzing Your Market 185
    9.3.1 The Promise of Unmet Needs/Wants 186
    9.3.2 Costs and Functions of Marketing 187
9.4 Pioneering StartUps 188
    9.4.1 Crossing the "Chasm" 189
    9.4.2 Your Pioneer Strategy 190
9.5 Setting the Right Price 190
    9.5.1 Establishing Your Pricing Tactics 191
    9.5.2 Tensions between Marketing and Sales 192
9.6 The Complex Sale 193
9.7 Market Segmentation 194
9.8 Positioning: Your Place in the Sun 195
    9.8.1 Famous Positioning Statements 196
    9.8.2 Template for a Positioning Statement 197
9.9 Sales and Sales Promotions 198
    9.9.1 Your Sales Force 198
    9.9.2 Sales Force Compensation 198
    9.9.3 Sales Force Time Management 200
References 201

10 Intellectual Property 203
10.1 Introduction 203
10.2 Patentable Inventions 206
    10.2.1 Who Benefits from Intellectual Property Rights? 207

10.3 Understanding Patents 207
    10.3.1 Contents of a Patent...208
    10.3.2 Claims: The Heart of a Patent 208
    10.3.3 Claims Categories...210
    10.3.4 Real Estate Analogy 211
10.4 Provisional Patents 212
    10.4.1 Time Limits...212
    10.4.2 Advantages and Disadvantages 213
    10.4.3 Patents Can Have Siblings/Children 214
    10.4.4 International Patent Organizations.............................214
10.5 Basis for Venture Investment 215
10.6 The America Invents Act..215
    10.6.1 First to File...216
    10.6.2 No Grace Period...216
10.7 The Special Case Involving Biotechnology Patents 218
    10.7.1 Genetic Engineering Patents......................................220
    10.7.2 Microbiological Sciences Patents 220
    10.7.3 Plant and Animal Sciences Patents............................220
    10.7.4 Pharmaceutical and Chemical Sciences Patents..........220
    10.7.5 Medical Sciences Patents..221
    10.7.6 Microorganisms and Sufficiency of Description 221
10.8 Practical Advice to Entrepreneurs/Inventors...............................221
    10.8.1 Who Is the "Inventor" in Your Company?...................222
    10.8.2 Conception 222
    10.8.3 Should You Out-License Your Patent? 223
10.9 Fundraising Importance of Patents...224
10.10 Trade Secrets in the StartUp Environment 225
References 226

11 Meet the JOBS Act 227
11.1 Introduction...227
11.2 The JOBS Act at a Glance...228
    11.2.1 Title III of the U.S. JOBS Act 229
    11.2.2 Equal Access and Disclosure 229
11.3 Crowdfunding 230
11.4 Issuer Requirements...231
11.5 Intermediary Requirements 231

11.6 Funding Portals 232
    11.6.1 Restrictions on Funding Portals 233
11.7 Mini-IPOs 234
11.8 Emerging Growth Company IPOs 234
    11.8.1 Relief for EGCs 235
    11.8.2 Confidential Filing Process 236
    11.8.3 Relaxed Restrictions on Investor Communications......237
References 238

12 Strategic Management 239
12.1 Introduction 239
12.2 Strategic Corporate Planning 240
12.3 Core Values and Core Purpose 242
12.4 Mission, Vision 242
12.5 Competitive Intelligence 244
12.6 Marketing Intelligence 246
    12.6.1 The Pizza Analogy 248
12.7 Strategic Marketing Goals by MI 250
References 252

13 Valuation Techniques 255
13.1 Introduction 255
13.2 Value vs. Price 256
13.3 Pre-Seed Financing 256
13.4 What Is Your Company Worth at StartUp (Seed Round)?.....257
    13.4.1 Initial Valuation at Seed Round 257
    13.4.2 Valuation at Series "A" Round 258
13.5 Valuation Approaches—Companies with Revenues 259
    13.5.1 Income Approach 260
    13.5.2 Discounted Future Returns Method 261
    13.5.3 Capitalized Returns Method 261
        13.5.3.1 Assumptions 262
13.6 Market Approach 263
13.7 Asset-Based or Cost Approach 263
13.8 Venture Capital Valuation 263
    13.8.1 Staged Capital Infusions 266
    13.8.2 Venture Capital Metrics 267
    13.8.3 The Investment "Hurdle Rate" 268
    13.8.4 Top Ten Lies VCs Tell 268

13.9   Vulture Capitalist 269
13.10   Last Words of Advice 270
        13.10.1 How to Maximize Your Company's Value 270
References 271

14 Writing a Winning Business Plan 273
14.1   Introduction..273
14.2   Need for a Winning Business Plan 274
14.3   The Essential Components 276
        14.3.1 Common Parts of a Good Business Plan 277
14.4   Risk Is a Four-Letter Word....................................277
14.5   Management Team That Can Execute.....................280
        14.5.1 Competent Team 281
14.6   How Innovative Is Your Invention?.......................282
14.7   Ignore Naysayers (and Prove Them Wrong) 282
14.8   Milestones Make Your Plan Real and Believable 284
14.9   The Value of Your Plan Will Be Measured by Its
        Implementation Potential 284
14.10   Example of a Winning Business Plan 285
References 285

15 Harvesting 335
15.1   Introduction..335
15.2   Start at the End...336
        15.2.1 Who Needs an Exit?...................................336
15.3   Capitalization Principles..337
        15.3.1 How to Capitalize Your StartUp 337
15.4   Going Public...339
        15.4.1 Advantages 339
        15.4.2 Disadvantages...340
        15.4.3 The Going Public Process.............................340
        15.4.4 The Registration Process.............................341
15.5   Strategic Alliances 343
15.6   The "Big Question" 344
        15.6.1 Drivers of M&A Activities 345
        15.6.2 Advantages of Strategic Alliances...................345
        15.6.3 Pitfalls of Strategic Alliances........................347
        15.6.4 Best Practices 347
15.7   Mergers and Acquisitions......................................350
        15.7.1 Mergers 350

15.7.2 Acquisitions 350

15.7.3 The Challenge of M&A 351

15.7.4 The Enduring Questions 352

15.8 The Day That Little David Acquired Giant Goliath—A Case
Study 352

15.8.1 The Acquisition 354

15.8.2 Open Heart Procedures 355

15.8.3 CardioTech as the White Knight 355

15.8.4 Risk Factors SWOT Analysis of the Gish Acquisition ... 357

15.8.4.1 Strengths 359

15.8.4.2 Weaknesses 359

15.8.4.3 Opportunities 360

15.8.4.4 Threats 360

15.8.4.5 The Stock-for-Stock Transaction 361

15.8.5 An Accretive Acquisition 361

15.9 Sample Prospectus 362

References 394

Index 395

About the Author 407

Acknowledgments

The author is indebted to Mark R. Szycher, MBA (General Motors, VP, Enterprise Risk Management) for his editing of several chapters, and to Scott R. Szycher, MBA (Membership & Marketing Manager, Associated Subcontractors of MA) for his co-authorship of Chapter 3.

Introduction

Welcome to *The Guide to Entrepreneurship: How to Create Wealth for Yourself and Your Stakeholders*. In contrast to academic publications that emphasize traditional teaching methods, this guide focuses on *you*, the entrepreneur, and how to become successful. "Your first priority is to become rich. Then, you can afford to become magnanimous."

At first glance, this *Guide* may appear to be a motivational/inspirational book masquerading as a business tome. That is how I truly feel about entrepreneurship. One of the transcendental moments in economic theory occurred in 1979 when David Birch published his seminal study on U.S. firms from 1969 through 1976 and found to his astonishment that small firms (<100 employees) accounted for 81% of new hires. His findings have been subsequently confirmed by other studies.

This *Guide* is arranged from the standpoint of the dominant founder or owner, and provides a how-to delineation of startup activities. To that end, I have highlighted the covered topics as

- Interpretation rather than description
- Practical advice rather than theory
- Referenced classical works to augment your basic understanding
- Startup activities and pitfalls
- The startup as a collective of managers (the "Dominant Coalition")
- The founder/entrepreneur as "manager of managers"

About This Book

"You don't have to be a meteorologist to know which way the wind is blowing."

The *Guide* demystifies the process of founding a firm from scratch. In writing this book, I have assumed that you are either (a) thinking of founding your firm or (b) have already decided to take the plunge. However, are you entrepreneurial timber? Do you have the leadership skills to raise all your needed capital? Do you dream of being your own boss and becoming rich in the process? If your answer is "yes" to these questions, then this book will provide you with a "how-to guide" to entrepreneurial success in plain English.

Although each chapter stands on its own, the book chapters are arranged in a sequential, temporal fashion. From *soup to nuts* as the old saying goes. In addition, as you progress from chapter to chapter, from an enterprise and personal standpoint, you should discern the following underlying structure:

Internal components	Strategy, tactics, mission, vision
External components	Networking, fundraising, alliance-building, strategic alliances, legal requirements

However, if you are an experienced entrepreneur, you can jump from chapter to chapter as you please, although I would encourage you to take a quick glance at the chapters as presented. You may find some unexpected golden nuggets.

God created the world in six days. What will you create today?

Chapter 1

Entrepreneurial Timber

1.1 Introduction

An **entrepreneur** is someone who establishes, organizes, manages, and assumes the financial risk of a new business venture. Entrepreneur comes from the French term *entreprende*, "to undertake or to build."

Entrepreneurship is the process of seizing an opportunity to introduce new products or services in an effort to transform inventions and innovations into monetary value. When entrepreneurship is used to describe activities within a firm or large organization, it is referred to as intrapreneurship and may include corporate venturing.

Entrepreneurs come from all backgrounds and operate within industries as diverse as construction, wholesale, retail, financial services, accounting services, manufacturing, drug development, medical device, transportation, and food service. Although individuals engaged in any of the above economic sectors are entrepreneurs in every respect, this Guide concentrates on characteristics of persons seeking to start new ventures that require a high degree of scientific knowledge.

An entrepreneur is typically a highly motivated individual possessing a clear vision to reach goals unimagined by the general population. He or she must have a high willingness to take risks ("If it fails, I will fix it"), display exceptional creativity (inventions and innovations), be fiercely independent (from formal supervision), can tolerate an atmosphere of continuous chaos, and craves controlling his or her own destiny.

In the U.S., small businesses are an important engine of economic growth. (The Small Business Administration, or SBA, defines a small business concern as one that is independently owned and operated, is organized for profit, and is not dominant in its field. Depending on the industry, size standard eligibility is based on the average number of employees for the preceding 12 months or on sales volume averaged over a 3-year period). Small businesses account for 95% of all business establishments, 50 to 60% of employment, and 60 to 70% of all new jobs.

Since the 1960s, thousands of new businesses have been started in the U.S., highlighting the promise of individual opportunity to other nations. The U.S. is a hotbed of entrepreneurial activity because:

■ Business failure is accepted by society without personal stigma.
■ Individual success is valued and cherished ("self-made man").
■ Change is anticipated and welcomed.
■ Intellectual property rights are afforded to small businesses.
■ Equity capital is available at all stages of venture formation.

Entrepreneurship flourishes despite discouraging statistics from the SBA: for every 100 small businesses started today, 24 will close within the first two years, 51 will fail in the first four years, and 60 will have filed for bankruptcy in six years. This is dramatically illustrated in Figure 1.1.

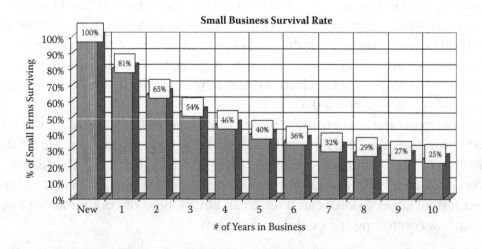

Figure 1.1   Small business survival ratio—On average, only 25% of small businesses survive by the tenth anniversary of their establishment.

1.2 Entrepreneurial Anatomy

Entrepreneurs have achieved celebrity status in many parts of the world. Who has not heard about Wal-Mart (from a few retail stores to the largest company in the world), or Hewlett Packard (two engineers in a garage with technical expertise but no product or plan creating a new industry), or Dell (from college dorm room to PC market leader), or Microsoft (brilliant youngsters without academic degrees outsmart giant IBM to dominate the world PC software market)?

Can we find common characteristics among these entrepreneurs? Scholarly research has identified a unique set of characteristics that are shared by all successful entrepreneurs. These characteristics are shown in Table 1.1.

1.3 The Entrepreneurial World

In the U.S., entrepreneurs start nearly 6 million small businesses annually, with approximately 11% of the adult U.S. population actively seeking to start a new business. Small businesses employ 51% of the private sector workforce, produce 51% of the nation's private GDP, and astonishingly create 24 times as many innovations than large companies, and, not surprisingly, generate 14 times as many patent applications per employee than large companies.

Table 1.1   The Entrepreneurial Mindset

1.	Exceptional leadership qualities
2.	Tolerance for risk, chaos, ambiguity, and uncertainty
3.	High creativity characteristics
4.	Overwhelming desire for achievement and peer recognition
5.	Ability to identify big market opportunities
6.	Willingness to accept and learn from failure
7.	Unshakable motivation to excel
8.	Fiercely independent
9.	Challenged by new opportunities
10.	Ability to create an organization supportive of their vision and mission
11.	Value achievement over money

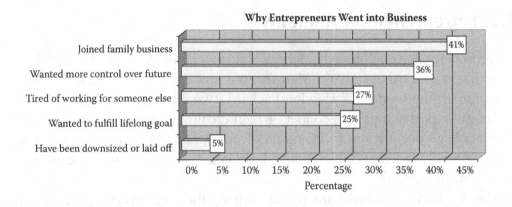

Figure 1.2 Why entrepreneurs went into business—Most entrepreneurs start their businesses to join a family business, or to better control their future.

Why do individuals enter the uncertain entrepreneurial world? Many will answer that "doing your own thing" and "doing what you like" is sufficient reward. Most agree, "If you do what you like, you will never work another day for the rest of your life." Figure 1.2 presents the five most common reasons why entrepreneurs go into business according to a recent Dun & Bradstreet survey of entrepreneurs.

Entrepreneurship is not confined to the U.S. Interesting statistics are offered online through http://gemconsortium.org/docs/download/2409. The Total Entrepreneurial Activity (TEA) provides a yearly snapshot of world activity, and is shown in Figure 1.3.

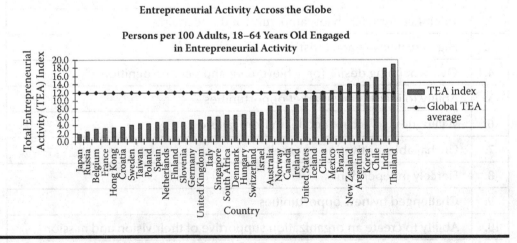

Figure 1.3 Entrepreneurial activity across the globe—India, China, Brazil, and Mexico are beehives of entrepreneurial activity when measured by the TEA index.

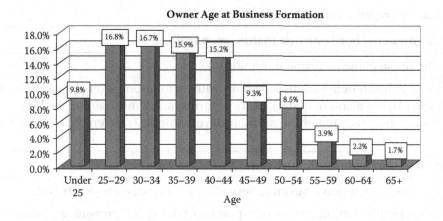

Figure 1.4   Owner age at business formation—The majority of entrepreneurs start their business between 25 and 44 years of age.

Another amazing characteristic of entrepreneurship is its diversity. Any person, regardless of age, race, gender, color, national origin, or cultural background, may become a successful entrepreneur. Let us consider age, for instance. Figure 1.4 shows the entrepreneur's age at new business formation according to a 2002 survey by the National Federation of Independent Businesses and Wells Fargo Bank.

1.4 Entrepreneurial FAQ

Many questions have been raised about what makes an entrepreneur tick. Table 1.2 and Table 1.3 present a summary of the traditional, most frequently asked questions by aspiring entrepreneurs.[1]

1.5 Success Characteristics

While it is very difficult to accurately pinpoint why some ventures fail and others thrive, certain personal characteristics of the founders seem predictive of success. Table 1.3 lists success characteristics of entrepreneurs. These eight characteristics are shown in Figure 1.5.

1.6 Failure Characteristics

There is rare unanimity among entrepreneurial scholars that the primary reason for failure is management incompetence.[2] Of course, "management incompetence" covers a lot of ground and can be a catchall phrase. Table 1.4 lists some of the most common mistakes made by "incompetent" management.

Table 1.2   Entrepreneurial FAQ

Are entrepreneurs born or made (nature vs. nurture)?
"Ride the horse in the direction it is going."
Successful entrepreneurs are usually born with an inherent ability in a given field (similar to playing piano or guitar), years of formal academic training, technical experience, contacts, know-how, and self-development. Thus, entrepreneurs are both born *and* made.
Can anyone start a business, anywhere, and at any time?
Fortune cookie: "To open a business very easy. To keep open very difficult."
The easiest part is starting; the hardest part is surviving and growing a venture that will create value for its founders and shareholders.
Are entrepreneurs gamblers or risk managers?
"A ship is safe while in harbor, but that is not what a ship is built for."
Entrepreneurs stack the odds in their favor by minimizing risks. Gamblers do not change the odds; they just play the game.
Entrepreneurs do not seek uncompensated risks, but conversely do not shy away from unavoidable risks.
Do entrepreneurs want to be the only star of the show?
"Half of this game is 90% mental."
Entrepreneurs have the vision, ability, and courage to start a new venture; therefore, for quite some time the limelight has been directly on them as individuals. Eventually, the entrepreneur needs to build a team, an organization, and a sustainable company; at that time, the whole stage lights up with all participants sharing in the applause.
Are entrepreneurs really their own bosses and completely independent?
"You can either complain that bushes have thorns—or rejoice that thorns have roses."
Ironically, the more successful the enterprise, the more "bosses" you will have. So much for "independence."
Entrepreneurs serve many masters and constituencies, including partners, investors, suppliers, creditors, bankers, lawyers, employees, advisors, consultants, distributors, and agents. These entities are collectively known as "stakeholders."

Table 1.2 Entrepreneurial FAQ (Continued)

Do entrepreneurs work harder and longer than comparable managers do?
"Commitment is like muscle; it is strengthened by use." Most entrepreneurs wish for a 9-day week. Time is their most implacable enemy. Starting a new venture is a lifestyle commitment; the entrepreneur must be willing to dedicate as much time as necessary to ensure success. Otherwise, it is a hobby. An entrepreneur is willing to work 16 hours a day for himself instead of only 8 hours for someone else.
Is entrepreneurship stressful and demanding?
"You must break eggs to make omelets." Starting a new business is stressful and demanding, but also highly rewarding. You will be faced with daily emergencies, personnel problems, governmental regulations, financial pressures, technical failures, etc. The list is endless. But this is your baby! It is like childbirth: you cannot do it alone, it is a bloody mess, and you will not sleep for months.
Is money the most important ingredient?
"Money is not everything, but it is way ahead of whatever is second." Undercapitalization is among the top three reasons for venture failures. Sixty percent of new businesses fail in the first three years primarily due to undercapitalization. However, money alone is not enough. The founder must also develop a good initial plan, execute, and follow up.
Do entrepreneurs need to understand financial statements?
"Waiting to be treated fairly because you are fair is like expecting the bull not to charge because you are a vegetarian." If you lack a basic understanding of financial statements, you will be treated unfairly. Know your numbers cold!
Are entrepreneurs always young and energetic?
"Only those who risk going too far will ever know how far they can go." According to the National Federation of Independent Businesses and Wells Fargo Bank survey conducted in 2002, 57% of all businesses were started by persons 35 years or older. Age brings experience and judgment. It is too bad that youth is wasted on the young.

(*Continued*)

Table 1.2   Entrepreneurial FAQ (Continued)

Do entrepreneurs seek unlimited power over others?
"Use your power to create respect and esteem." By virtue of their accomplishments, entrepreneurs *earn* power. Power is conferred, not taken. Power is a by-product of achievement, and only grows with time.
If an entrepreneur is sufficiently talented, is success assured in a couple of years?
"Lemons ripen in two years, but pearls take seven or eight years." (old venture capital maxim) Most businesses are lucky to break even in 4 to 5 years from startup.
Can anyone with just a good idea be funded?
"Ideas are a dime a dozen. Have you proven the concept?" While great projects always start with good ideas, ideas alone rarely raise money. The entrepreneur must put "skin in the game" by self-funding the concept.
If an entrepreneur is well funded, is he or she invincible?
"A fool and his money are soon parted." Too much initial capital may well be a detriment as the founder may suffer too much dilution and may start acting like a spoiled brat.
Are entrepreneurs lone wolves that cannot work with others?
"Entrepreneurs are like Formula 1 race-car drivers: they should not forget about their pit crew." Successful entrepreneurs are skilled at selecting, training, and nurturing a great team. No lone wolves here—backing from a great team will lower risks.
If you fail in one venture, can you ever raise funds again?
"When God closes a door, he opens a window." Businesses fail, but entrepreneurs do not. Failure is often the fire that tempers the entrepreneurial metal. A lack of fear of failing is a distinguishing entrepreneurial trait. Failure is a station on the way to success. Successful entrepreneurs accept failure *professionally.*
Are entrepreneurs solely motivated by financial rewards?
"If we are so rich on paper, why are we still driving a Pinto?" Entrepreneurs are more motivated by the opportunity to create their own world, do what they enjoy most, and desire recognition of their accomplishments more so than realizing long-term capital. Nonetheless, because wealth is actively sought in our society, achieving multi-millionaire (or multi-billionaire) status doesn't hurt either.

> **Characteristics**
>
> ◆ High achiever
> ◆ Risk manager
> ◆ Problem solver
> ◆ Status seeker
> ◆ High energy level
> ◆ Self confident
> ◆ Maximum responsibility and authority
> ◆ Craves personal satisfaction

Figure 1.5 Characteristics—The eight characteristics that predict success in business founders.

Table 1.3 Success Characteristics

High achiever	Inexhaustible commitment and enthusiasm. Maxim: Create like a god; command like a king; work like a slave.
Risk manager	The entrepreneurial equation: Value = Rewards/Risks The name of the game is to lower your technical, managerial, marketing, and regulatory risks for your investors.
Problem solver	Edison did not invent the light bulb. He solved the problem that 22 others who preceded him failed to do.
Status seeker	"It is easier to go down a hill than up, but the best view is from the top." Be the undisputable leader and visionary.
High energy level	"Success is largely a matter of hanging on after others have quit." Hang in there until you succeed.
Self confident	"If you don't believe in yourself, who will?" Sell yourself first, then the venture.
Maximum responsibility and authority	"If there is no wind, row." Only you can make the venture succeed. Develop a high tolerance for possible failures.
Craves personal satisfaction	"Only dead fish swim with the current." Lead and you will be personally rewarded.

Table 1.4 Failure Characteristics of "Management Incompetence"

Underfunding	Accepting dangerously low initial funding in hopes of reaching milestones that will bring additional cash in the future.
Inability to raise follow-up funding	A corollary of underfunding. Creating dissatisfied investors.
Cash flow management	Cash is king. Run out of cash and you will run out of luck.
Extravagant operating expenses	The "Taj Mahal" complex. Trying to impress investors with the latest gadgets, overly impressive headquarters, expensive shiny brochures, etc.
Overestimating your true market size	This is a crucial marketing research failure. One of the top ten mistakes made by inexperienced founders.
Ignoring your competition	"Keep your friends close, and your enemies even closer."
Underestimating your development costs	Remember the old maxim: "It will cost you twice as much and take you twice as long as you thought."
Inability to comply with regulations	Regulations never get easier; they only get tougher.
No real competitive advantage	
Poor sales forecasts	Without sales, there is nothing to manage. Be realistic about your sales projections and "make your numbers."
Wrong business model	A business model incorporates all the activities that define how a firm will compete in the marketplace. What is your value chain?
Not understanding customers' needs	Do your product features solve customers' problems? If not, your product is merely a "Me, too."

1.7 Risk Is a Four-Letter Word

In general, risk is the probability of loss or damage. In an entrepreneurial sense, risk is the chance of losing invested capital. Thus, in a new venture, risk is the very real possibility that the business will fail or be less successful than projected.

Peter Drucker, the management guru, identified four types of business risks:[3, 4]

1. Risk that is built into the very nature of the business and cannot be avoided
2. Risk one can afford to take
3. Risk one cannot afford to take
4. Risk one cannot afford not to take

Generally, there are six major sources of risk in an entrepreneurial venture as summarized in Table 1.5.

The aspiring entrepreneur must be willing to live (and thrive) in an atmosphere replete with risk, which is one of the distinguishing characteristics of entrepreneurs, however, not with exaggerated uncontrollable, undefined risks. The founder must be a risk manager, not a gambler (where the odds are against the player).

1.8 Challenges of Entrepreneurship

Before you lose your money, time, energy, family, and possibly your health in a new business venture, carefully consider the crucial question shown in Figure 1.6.

Table 1.5 Major Sources of Risk

Technological	Does the new technology work?
Financial	Can you raise enough capital?
Market	Is the market large enough?
Regulatory	Can you meet FDA, EPA, OSHA, ISO, SEC, IRS etc. regulations?
Operational	Is adequate management in place?
Force Majeure	Are you prepared for natural disasters, fire, floods, hurricanes, etc.? Can you afford adequate business insurance?

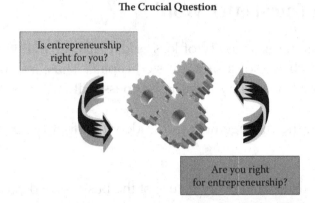

Figure 1.6   The crucial question—Are you entrepreneurial timber?

This Guide will help you assess your chances of success by maximizing your efforts, gaining confidence in your own abilities, and evaluating the rewards of being your own boss. Ask yourself the following questions as shown in Table 1.6:[5]

- Are you considering starting your own venture?
- Are you a corporate executive ready to quit your job and strike out on your own?
- Are you a corporate executive willing to start a corporate-backed venture (intrapreneur)?
- Are you a highly skilled person in a fast-growing technical field?
- Have you been downsized, laid-off, or re-engineered out of your job?
- Are you driven to create maximum value?
- Are you pursuing a fast-track entry into entrepreneurship (acquisition)?

Ironically, if you decide to start your new venture, your next important task is to decide how you will exit once the venture has succeeded. Exit is the means by which investors (and founders) dispose of their investments. Your exit strategy will determine whether capitalists will invest in your company, or by-pass it altogether (Table 1.6).

1.9 The Ten Commandments of Entrepreneurship

We have now reached the point where you have at least superficially completed the process of *objectively* considering the commitment of conceptualizing, implementing, financing, and operating your new business. Figres 1.8

Table 1.6   Exit Strategies

Exit Strategies	New Owners
Initial public offering	Public shareholders
Acquisition	Third-party (a strategic or financial buyer)
Secondary sale out-licensing	Third-party and entrepreneur
Buyback	Entrepreneur and management
Liquidation	Bankruptcy Putting the company to sleep

and 1.9 are "Rules for entrepreneur, I and II." Subsequent chapters will help you to more rigorously explore all the facets involved in your new venture.

Good decision-making on your part will require a focused approach to both technology and business. Moreover, as Winston Churchill famously said, it will require "a lot of blood, sweat and tears." However, the rewards can be astronomical.

Figure 1.11 are the Ten Commandments of Entrepreneurship. You have started a great adventure with a one-way ticket. Bon voyage!

Figure 1.7   The great irony—Start at the end (decide your exit strategy).

Rules for Entrepreneurs I

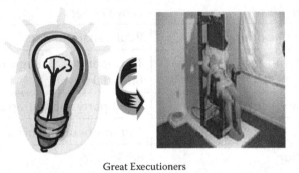

Great Executioners

Figure 1.8   Rules for entrepreneurs I—Be a great executioner.

Rules for Entrepreneurs II

Team   Customers/investors

Great Communicators

Figure 1.9   Rules for entrepreneurs II—Be a great communicator.

Ten Commandments of Entrepreneurship

- Thou shall be a leader first, then a manager
- Thou shall take risks, not gambles
- Thou shall find a screaming problem then solve it
- Remember that time is your worst enemy
- Honor your best achievers
- Thou shall focus on creating value
- Thou shall think big, then execute
- Thou shall under-promise and over-deliver
- Thou shall be your venture's best salesperson
- Thou shall not develop solutions looking for a problem

Figure 1.10   Ten commandments of entrepreneurship—Memorize and accomplish to succeed.

References

1. Modified after Timmons, J.A. and Spinellis, S., *New Venture Creation*, 8th ed., New York: McGraw Hill Companies, 2009, Ch. 2.
2. Dun & Bradstreet, *Business Failure Record*, New York: Dun & Bradstreet, Short Hills, NJ, 1995.
3. Drucker, P.F., *Innovations and Entrepreneurship*, HarperBusiness, New York, reissue edition, 2006.
4. Drucker, P.F., *Managing for Results*, HarperBusiness, New York, reissue edition, 2006.
5. Adapted from Roth, C., *The Entrepreneur Equation*, Dallas, TX: BenBella Books, Inc., 2001.

Chapter 2

Intrapreneurship: Corporate Entrepreneurship

2.1 Introduction

"Sooner or later your competitors will slow your sales, and profits will begin to evaporate. The solution is Intrapreneurship."

Some of the world's most successful companies, including 3M, Anaconda-Ericsson, GE, Lockheed, Intel, Apple Computers, Rubbermaid, Google, Sony, Toyota, Thermo Electron, Medtronic, and IBM, are all enthusiastic supporters of intrapreneurship. Could that be their secret of success?

2.2 Definitions

An **intrapreneur** is an innovator within an existing, large company. In their influential 1978 publication, Gifford Pinchot III and Elizabeth Pinchot[1] coined the phrase "intrapreneurship" to describe the marriage of an entrepreneurial spirit—complete with its fierce independence and lack of deference to established views and the confines of conventional wisdom—with the resources of a large corporation. Later, Norman Macrae credited the term to Gifford Pinchot III in the April 17, 1982, issue of *The Economist*.[2]

In 1992, the American Heritage Dictionary officialized the popular use of intrapreneur to mean, "A person within a large corporation who takes direct responsibility for turning an idea into a profitable finished product through assertive risk-taking and innovation." Intrapreneurship is now known as practicing a corporate management style that integrates risk-taking and innovation with reward and motivational techniques more traditionally associated with entrepreneurship.

In 1994, Pinchot and Pinchot published their "Intraprise Manifesto,"[3] which is summarized next:

- Employees are treated with independence, dignity, and responsibility.
- Employees make up their own minds about what to do, limited only by general rules and commitments.
- Employees have access to resources required by their ideas and are held accountable.
- Employees have the authority to manage the resources of their business units without interference.
- Employees are free to select their associations (e.g., peer support, coaches).
- Employees are permitted, and even rewarded, for taking qualified risks and making mistakes.

Using the failed examples of Wang Computers, DEC, Polaroid, and Kodak, Investopedia[4] suggests that companies should encourage employees to explore ideas rather than waiting until the company is in a bind. If an idea looks profitable, the company provides the innovator an opportunity to become an intrapreneur. Intrapreneurs thus use their entrepreneurial skills without personally incurring financial risks associated with entrepreneurial activities.

2.2.1 Kelly Johnson's 14 Rules of Skunk Works[5]

Johnson earned recognition for his contributions to noteworthy aircraft designs including the Lockheed U-2, SR-71 Blackbird spy planes, P-38 Lightning, P-80 Shooting Star, and the F-104 Starfighter. Johnson's famed "down-to-brass-tacks" management style was summed up by his motto, "Be quick, be quiet, and be on time." He ran Lockheed's Skunk Works using "Kelly's 14 Rules." (Note: Skunk Works is a term used to describe keeping a group of intrapreneurs separate from the rest of the firm.)

1. The Skunk Works manager must be delegated near complete control of all aspects of his program, and should report to a division president or higher.
2. Strong but small project offices must be established by both the military and industry.
3. The number of people having any connection with the project must be strictly limited. Use a small number, 10% to 25% of a typical project's staffing of top-notch people.
4. A very simple drawing and drawing release system with great flexibility for making changes must be provided.
5. Require a small number of reports, but important work must be recorded thoroughly.
6. There must be a monthly cost review covering not only what has been spent and committed but also what expenditures are expected to complete the program. Don't prepare the books 90 days late, and don't surprise the customer with unexpected cost overruns.
7. The contractor must be delegated and must assume more than normal responsibility to get good vendor bids for subcontract on the project. Commercial bid procedures are very often better than military ones.
8. The current inspection system used by the Skunk Works, approved by both the Air Force and Navy, meets the intent of existing military requirements and should be used on new projects. Push more basic inspection responsibility back to sub-contractors and vendors, and don't duplicate such inspection.
9. The contractor must be delegated the authority to test his final product in flight. He can and must test it in the initial stages. If not, he rapidly loses his competency to design other vehicles.
10. The specifications applying to the hardware must be agreed to well in advance of contracting. The Skunk Works' practice of having a specification section stating clearly which important military specification items will not knowingly be complied with and reasons therefore is highly recommended.
11. Funding a program must be timely so that the contractor does not have to keep running to the bank to support government projects.
12. There must be mutual trust between the military project organization and the contractor with very close cooperation and liaison on a day-to-day basis. This cuts down misunderstanding and correspondence to an absolute minimum.
13. Access by outsiders to the project and its personnel must be strictly controlled by appropriate security measures.

14. Because only a few people will be used in engineering and most other areas, ways must be provided to reward good performance by pay not based on the number of personnel supervised.

2.3 Theory of Intrapreneurship

"Lead, follow, or get out of the way."

As we have seen, an entrepreneur is an independent person who starts a venture, bears the full risk of failure, and enjoys the complete fruit of success, whereas an intrapreneur is partially independent and is sponsored by the corporation in which he or she works. Intrapreneurs are not liable for financial losses in case of failure (although they may risk dismissal). An entrepreneur raises the finance from various sources and promises a substantial return, whereas an intrapreneur does not assume responsibility to raise capital or to return it. An entrepreneur has no relation with any existing organization, whereas intrapreneurs operate within the organization where they work, as shown in Table 2.1.

Intrapreneurship involves vision, innovation, risk-taking, and creativity. Intrapreneurs imagine things in novel ways. Intrapreneurs have the capacity to take calculated risks and accept failure as a learning point. An intrapreneur thinks like an entrepreneur in seeking those opportunities that will ultimately benefit the organization. Intrapreneurship is a novel way of making organizations more profitable by incubating imaginative employees' entrepreneurial thoughts. It is in the best interest of an organization to encourage intrapreneurs because it is an effective way for large companies to re-invent themselves and improve financial performance.

Table 2.1   Significant Differences between Entrepreneurs and Intrapreneurs

Entrepreneur	Intrapreneur
Independent agent	Sponsored by employer
Risks own capital	Not financially liable for failures
Raises capital (either equity or debt)	Financed by employer
Establishes a new entity	Operates within an existing, large organization

Large corporations such as IBM, General Mills, AT&T, and Apple have been amply rewarded by the market for their intrapreneurial prowess, although the ubiquitous Post-it® Notes has become legendary in the annals of intrapreneurship. 3M's policy of allowing employees to use 15% of their time and resource on pet projects, known as "bootleg time,"[6] is more fully discussed in the following paragraphs.

2.3.1 *The 3M Illustration*

The Post-it® Note was originally a solution looking for a problem. In 1972, Dr. Spencer F. Silver patented (U.S. patent 3,691,140) a unique paper adhesive that permitted easy removal without tearing, and permitted re-bonding without application of additional adhesive; ironically, the 3M scientist didn't know what to do with his discovery. Six years later, a colleague of Dr. Silver, Art Fry, remembered the light adhesive when he was daydreaming about a bookmark that would stay put in his church hymnal. While the intrapreneur did not receive any profits from his invention, he was rewarded with a promotion that included a pay raise and many company benefits.

A Post-it® Note (or Sticky Note) is stationery with a re-adherable strip of adhesive on the back designed for temporarily attaching notes to documents and other surfaces. In 1968, Dr. Spencer Silver, a chemist at 3M in the United States, was attempting to develop a super-strong adhesive; instead, he accidentally created a "low-tack," reusable, pressure-sensitive adhesive. For five years, Silver promoted his invention within 3M, both informally and through seminars, without much success. In 1974, a colleague of Silver's, Art Fry, who had attended one of Silver's seminars, used the adhesive to anchor his bookmark in his hymnbook. Fry then further developed the idea by taking advantage of 3M's officially sanctioned "permitted bootlegging" policy. 3M launched the product in stores in 1977 in four cities under the name "Press 'n Peel," but its results were disappointing. A year later, 3M issued free samples to residents of Boise, Idaho, and 94% of the people who tried them said that they would buy the product. On April 6, 1980, the product debuted in U.S. stores as "Post-it Notes." In 1981, Post-its® were launched in Canada and Europe.

There were actually two accidents that led to the invention of the Post-it® Note. The first was Spencer Silver's creation of the Post-it's adhesive. According to the former Vice President of Technical Operations for 3M Geoff Nicholson (now retired), in 1968, Silver was working at 3M trying to create super-strong adhesives for use in the aerospace industry in building planes. Instead of a super-strong adhesive, though, he accidentally created an incredibly weak, pressure-sensitive adhesive agent called Acrylate Copolymer Microspheres.

This adhesive did not interest 3M management, as it was seen as too weak to be useful, but nonetheless had two interesting features. First, when stuck to a surface, the adhesive could be peeled away without leaving any residue. Specifically, acrylic spheres only stick well to surfaces where they are tangent to the surface, thus allowing weak adhesion to be peeled easily. Second, the adhesive is re-usable, thanks to the spheres' incredible strength and resistance to breaking, dissolving, or melting. Despite these notable features, no one, not even Silver himself, could devise a good marketable use. Thus, even with Silver's promotion to various 3M employees for 5 years, the adhesive was shelved.

Finally, in 1973, when Geoff Nicholson was named 3M's Products Laboratory Manager, Silver approached him immediately with the adhesive and provided samples to play with. Silver also suggested his best idea for adhesive: a bulletin board with the adhesive sprayed on. One could then stick pieces of paper to the bulletin board without tacks, tape, or the like; the paper subsequently could be easily removed without any residue left on the sheets. While a decent idea, the sticky bulletin board was not seen by Nicholson as sufficiently profitable, principally because annual bulletin board sales are low.

Enter the second accident, this one courtesy of chemical engineer Arthur Fry. Fry was a 3M Product Development Engineer, and was familiar with Silver's adhesive thanks to attending one of Silver's seminars. Fry sung in a church choir in St. Paul, Minnesota, and while he sang, the page marker kept falling out of the hymnal. Fry eventually had a stroke of genius—he used Silver's adhesive to keep the slips of paper in the hymnal. Fry then suggested to Nicholson and Silver that they were using the adhesive backwards. That is, instead of sticking the adhesive to the bulletin board, they should "put it on a piece of paper and then we can stick it to anything," Fry exclaimed.

Doing so proved easier said than done, in terms of practical application. Although it was easy enough to get the adhesive onto the paper, the adhesive would often detach from the paper and stay on the object the paper was stuck to. Silver's bulletin boards did not suffer from the problem as he fabricated the boards to bond better with the board than with the paper. Two other 3M employees now entered the scene: Roger Merrill and Henry Courtney. Merrill and Courtney were tasked with devising a coating that would remain on the paper and not on whatever the paper was stuck to when the paper was removed; the two engineers ultimately succeeded.

Nonetheless, since 3M's management still did not think the product would be commercially successful, the idea was more or less shelved for three years; ironically, Post-it® Notes were extremely popular internally at 3M labs during that time. Finally, in 1977, 3M began running test sales of the Post-it® Note in four cities; at that time, it was called "Press 'n Peel." The test sales were again disappointing, which confirmed in the executives' views that the product was not commercially viable.

Fortunately for offices the world over, Nicholson and Joe Ramey, Nicholson's boss, didn't give up; they felt the marketing department had dropped the ball, as businesses and individuals were not provided a sufficient number of samples to see for themselves how useful the notes could be. A year after the initial flop, 3M introduced the Post-it® Note to the world, providing a huge number of free samples in Boise, Idaho, in a marketing campaign labeled "The Boise Blitz." This time, the re-order rate went from almost nil in the previous attempt to 90% of the people and businesses that had received the free samples. For reference, a 90% re-order rate was double the best initial rate 3M had ever achieved for any other product introduction. Two years later, the Post-it® Note was marketed throughout the U.S.

Thus, after 5 years of rejection and another 7 years in development and initial marketing failures, Sliver and 3M finally had a hit in Post-it® Notes, which have since become a mainstay in offices the world over. In fact, Post-it® Notes remains one of the top five selling office supply products in the world,[7, 8] and 25 years later, Post-it® Notes is still one of 3M's top-grossing products.

Over the years, 3M has had a policy that at least 25% of its revenue should come from products introduced within the last 5 years. That encourages a culture of innovation and intrapreneurship.

Table 2.2 summarizes 3M's rules for innovators.

Large companies may encourage intrapreneurship by:

- providing encouragement to innovators
- tolerating failures
- reducing the need for periodic explanations
- circumventing procedures, red tape, and micromanaging
- targeting real market needs
- providing "innovation time" to research, discuss, and execute projects
- rewarding successful intrapreneurs in ways visible to all employees

Table 2.2   3M's Rules for "Bootlegging"

• Don't kill a project
• Tolerate failure
• Keep divisions small
• Motivate the champions
• Stay close to the customer
• Share the wealth

Table 2.3 presents some important strategies that encourage an intrapreneurial activity.

2.4 Characteristics of Intrapreneurs

Only one animal in the world is willing to endure peer ridicule, work *pro bono*, and face continual rejection and prolonged mental anguish in the hope of making a great contribution to his or her company: the

Table 2.3   Strategies that Encourage Intrapreneurial Activity

Support idea generation. Grant project ownership.	"Make intrapreneurial heroes out of managerial robots."
Make risk-taking and failure non-fatal.	"Failure is a station on the way to success."
Train employees on creating and selling innovation.	"Intrapreneurs are managers of managers."
Celebrate and reward intrapreneurial behavior.	"To be a realist you must be a dreamer."
Encourage networking and collaboration.	"Peer recognition is more important than salary, bonuses, or promotions."
Provide access to data and business intelligence.	"Keep your friends close, and your enemies even closer."
Contain the stifling bureaucracy.	"Size is the enemy of innovation."
Create a common fund for intrapreneurial initiatives.	"Raising cheap money is very expensive."

Modified after "10 tips for turning employees into 'intrapreneurs,'" SmartPlanet http://www.smartplanet.com/blog/bulletin/10-tips-for-turning-employees-into-intrapreneurs/, accessed March 2013.

Table 2.4   Recognized Characteristics of Intrapreneurs

"Lions don't need to roar."
• Goal oriented, impactful, focused
• Ambitious, brilliant
• Competitive, activist for change
• Questioning, defiant, ready to do battle
• Self motivated, inspiring
• Spurns bureaucracy, rebellious
• Comfortable with change, determined
• Adept at internal politics
• Good at conflict resolution
• Able to lead others

"intrapreneur." Intrapreneurs are pioneers with full knowledge that "pioneers have all the tomahawks in their backs."

Table 2.4 summarizes important characteristics of intrapreneurs.[9]

Corporate entrepreneurship is a constellation of processes utilized by existing large organizations to create new business opportunities. Individuals who perform these services are known as corporate entrepreneurs, intra-corporate entrepreneurs, or intrapreneurs. In this chapter, we will refer to corporate entrepreneurs, intra-corporate entrepreneurs, or intrapreneurs interchangeably.

2.5 External and Internal Business Environment

The business environment is the totality of factors affecting an ongoing business, consisting of the external and internal environments. The external environment encompasses competitive trends, government regulations, and economic factors, whereas the internal environment relates to all organizational activities under the direct control of the business, as summarized in Figure 2.1.

- Economic factors: overall health and vitality of the economic environment in which the organization operates
- Technological factors: proprietary techniques to convert resources into valuable products or services

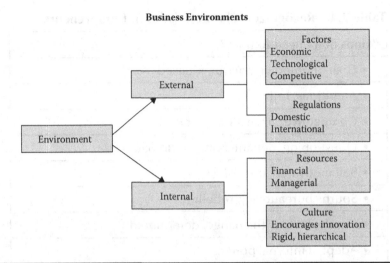

Figure 2.1   Business environments—Intrapreneurs should consider both external and internal factors likely to affect their proposed business.

- Regulations: all governmental and trade constraints imposed upon the business
- Financial resources: monetary funds available to a business for spending in the form of cash, liquid securities, and credit lines
- Managerial resources: portfolio of capabilities and strategies to create maximum value for the organization
- Organizational culture: constellation of values, beliefs, behaviors, customs, attitudes, and history that guides members to understand what is expected and what the organization considers important ("How we do things around here.")

Business environments are constantly changing, and a business capacity to adapt, to innovate, and to respond to competitive pressures may be directly impacted by intrapreneurs. If given the chance, intrapreneurs can lower production costs, improve quality, create new products, and open new markets; but to do so, large companies need to establish the right culture, which will be explored in detail in the next section.

2.6 Culture Regarding Innovation

Innovation is taking a creative idea and providing benefits to the user. Innovation is an elusive activity and a never-ending task, especially because strategic objectives are constantly changing. Intrapreneurs can openly

Table 2.5   A Clash of Cultures

"Be a coach, not a judge."	
Corporate Culture	*Intrapreneurial (Innovation) Culture*
Rewards ultraconservative decisions	Trial and error "You will miss 100% of the shots you don't take."
Demands to wait for instructions "It usually takes 3 weeks to prepare a good impromptu speech."	Rewards quick actions "Don't punish failure; reward success."
Expects "no surprises"	Encourages new approaches that may fail "Starting up is hard to do."
Collects information "Paralysis of the analysis."	Expects decisions even under imperfect information "Take risks, not chances."
Controls information "Information is power."	Encourages open discussion "Gentlemen do read each other's mail."

explore, express, and undertake their creative insights in a culture of innovation. Therefore, management is challenged with creating incentives that are consistent with an innovation-friendly culture. Table 2.5 presents the primary differences between traditional corporate and intrapreneurial cultures.

In an established company, the intrapreneur can be a *change agent*, providing a much-needed competitive advantage. This is depicted in Figure 2.2.

2.7 Corporate Support for Internal Business Creation

"Don't fund the problem; fund the solution."

Surprisingly, the biggest barrier to intrapreneurship is corporate culture. **Culture** is the set of shared attitudes, values, goals, practices, and expectations that characterize an organization. Culture is the crucial intrapreneurial ingredient. Without a corporate culture supportive of internal innovation, there can be no intrapreneurship.

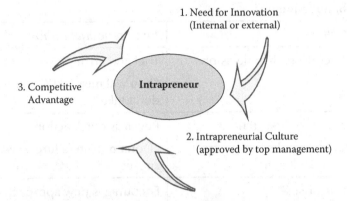

Figure 2.2   Central role of the intrapreneur—As change agent, the intrapreneur should clearly articulate the need, culture, and competitive advantage of the proposed business.

An intrapreneurial culture:

1. identifies sources of opportunities,
2. codifies the process of initiation, development, and introduction of opportunities, and
3. selects individuals most likely to discover, evaluate, and exploit opportunities.

Top management can encourage intrapreneurship in many ways. Table 2.6 provides a summary of some established methods of encouraging and fostering intrapreneurship.

2.8 Critical Issues in Intrapreneurship

"From know-how to innovation."

Intrapreneurship within a large organization starts with an idea and ends with commercial application. An innovation must be new, better, cost effective, and solve a screaming need in the market. The intellectual process of transforming an idea into know-how can be summarized by the acronym DIKK (Data, Information, Knowledge, and Know-how), as shown in Figure 2.3.

Table 2.6   Encouraging Intrapreneurship

"Uncertainty makes innovation less expensive."		
Formal steps that lead to the creation of a more innovative culture within an existing corporate organization		
Parameters	*Measures*	*Factors*
Regular compensation, job security, innovation time	Personal rewards Specific goals	Promotions, autonomy, peer recognition, bonuses, empowerment
High-level support Executive encouragement	Top management buy-in under corporate umbrella	Public commitment, space, semi-autonomy
Materials, budgets, data analysis, attendance at trade meetings Competitive intelligence	Establish well-defined project resources Feedback and positive reinforcement	Know-how, knowledge, access to competitive intelligence
Hierarchy	Clearly defined organizational chart Formal control structure	Skunk Works, brainstorming, cross-functional teams
Tolerance of failure	Risks, grieving period Support system	No penalties or dismissal

Approach to solving problems:

- **Data:** Individual elements.
- **Information**: Categorization, summarization, classification.
- **Knowledge**: Analysis, comprehension, relationships, acquaintance with the facts and information. You know it, *in your head*.
- **Know-How:** Competitive intelligence; ability to perform a task or action in a highly competitive environment.

Table 2.7   Data vs. IKK

Data	*IKK*
Parts per hour manufacturing or assembly	Maximizing profits
Receiving governmental approvals	Entering new market; new product launch
Reducing device-associated infections	Greater market share

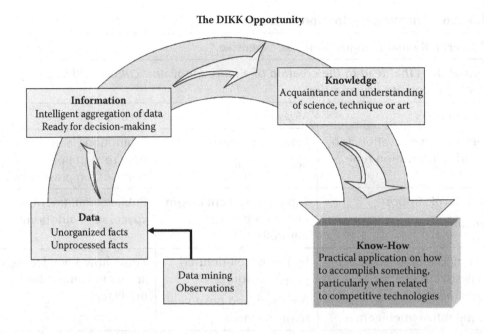

Figure 2.3   The DIKK opportunity—Data, information, knowledge, and know-how (DIKK).

In plain English, information is what you store in your desk; knowledge is what you carry in your head. While everyone else is drowning in a sea of information, the intrapreneur uses knowledge as a life raft. The process of innovation generally starts with a single visionary, as shown in Figure 2.4.

2.9 The Spin Zone

"Killing two stones with one bird."

There are two types of corporate venturing: external and internal. *External corporate venturing* is the creation of a semi-autonomous entity, known as a spin-out, separate from the parent organization. External ventures, known as spin-offs, also include divestitures, that is, selling assets, divisions, or subsidiaries to another corporation, a combination of corporations, or an individual. A variation on the spin-off is an equity carve-out.

Internal corporate venturing is the creation of an entity residing within the existing organization, for example, a new division, and is often run by managers-turned-intrapreneurs.

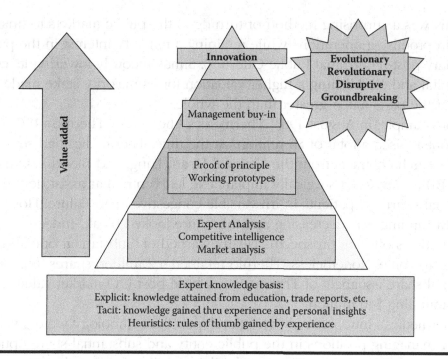

Figure 2.4   The innovation process—You as the innovator.

Clearly, a spin-out is not the same as a spin-off, as spin-out is a type of spin-off whereby a company "splits" a section to be operated as a separate business. The common definition of spin-out is when a division of a company or organization becomes an independent business. The "spun-out" entity takes assets, intellectual property, technology, or existing products from the parent organization, but remains firmly within the *corporate umbrella.*

2.9.1 The Spin-Out

"Making the children behave."

The poster child of the spin-out technique was Thermo Electron ("Thermo"), now known as ThermoFisher Scientific.[10] While most companies "spin-off" their failures, Thermo demonstrated in the early 1980s how to "spin-out" their successes. Thermo contributed seed capital, but when a new business started requiring significant capital, the venture was permitted to "go public." At the time Thermo sold minority shares to the public in one of its emerging businesses, those operations were marginally profitable or running at an operating loss. However, under the Thermo corporate umbrella and reputation, the fledgling company commanded an enviable initial valuation.

This was a surprising method of turning to the public markets to finance certain promising operations while retaining a majority interest in the public entity post-IPO (Initial Public Offering). Simultaneously lowering its cost of capital and establishing a higher valuation for its majority stake made the sum of the parts much greater than the whole.

For example, in August 1983, Thermedics (the first of Thermo's IPO spin-outs) raised a total of $6 million. At the time, Thermedics had several research contracts from the National Heart, Lung, and Blood Institute (NHLBI) to develop a surgically implantable left ventricular assist device (artificial heart) for patients in irreversible congestive heart failure. However, NHLBI funding was decreasing and the future looked bleak. Indeed, Thermedics's offering prospectus clearly warned of "substantial operating losses" to come. Nonetheless, Thermo retained 4.2 million shares (86.6% of the total share issuance) of Thermedics, whose post-IPO market value was an astounding $40 million.

Thermedics's intrapreneurs who developed the technology were given senior managing positions in the public entity and substantial share options, thus bonding them to the Thermo "family" and discouraging them from leaving to form competing firms. Thermo's culture encouraged intrapreneurship by helping intrapreneurs take their divisions public. If an employee had innovative products and intrapreneurial drive to create his or her own business, Thermo would be there to help.

Table 2.8 summarizes the spin-outs engineered by Thermo between the years 1983 and 1995.[11]

For every new spin-out technology, Thermo created an intrapreneur's paradise by promoting the innovators with administrative, financial, legal, marketing, and organizational backing. The intrapreneurs and technical talent of the public companies gained autonomy, stock ownership, and control without adult supervision. Investors bought into a clear and focused venture whose strategic and financial potential was not diluted by the labyrinthine hierarchy of the parent organization. Thermo was white hot.

At the time of the 12 spin-outs, Thermo was still being run by the original founders, the Hatsopoulos brothers, who were not afraid of being outshined by their intrapreneurs or losing their jobs to a younger crowd. Perhaps professional managers would be worried about losing control or their power base.

From 1993 to 1996, Thermo stock price tripled. However, by 1998, equity analysts were confused by so many "Thermo children and grandchildren." Each child had its own board, products, and target market. Businesses

Table 2.8   Spin-Out Culture Thermo "Children" and "Grandchildren"

Spin-Out Company Name	IPO Date Month/Year	IPO Price per Share ($)
Thermedics	8/83	9.50
Thermo Instrument	8/86	8.00
Thermo Process	8/86	6.00
Thermo Power	6/87	8.50
Thermo CardioSystems	1/89	8.50
Thermo Voltek	3/90	1.12
Thermo Trex	7/91	12.00
Thermo Fibertek	11/92	8.00
Thermo Remediation	12/93	12.50
ThermoLase	7/94	6.00
Thermo Ecotek	2/95	12.75
ThermoSpectra	8/95	14.00

overlapped and customers became confused—Thermo had gone too far with this model. Many of the spin-outs contained core businesses that should have remained within the parent, while others were non-core businesses that should have been completely spun-off by traditional methods. In 1998, Thermo stock lost $40 per share in a single year, hitting bottom at $10 per share. By the end of 2001, all the spin-outs were stopped. Their early success had turned into late excess.

Notwithstanding, the Thermo spin-out model still remains a viable alternative for many large companies wishing to reward divisional intrapreneurs who aspire to become independent businesspeople.

2.9.2 The Spin-Off Reorganization

"Every battle is won before it is fought."

ATT/Lucent, GM/Delphi, DuPont/Conoco, and RJReynolds/Targacept did not follow the Thermo model. Instead, they followed the spin-off model. The parents granted outright ownership of all intellectual property and divested themselves from controlling interests,[12] as shown in Figure 2.5.

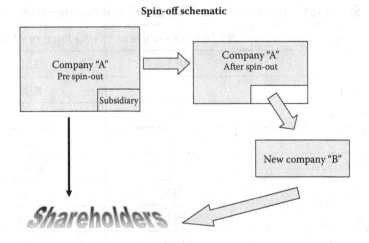

(1) Shareholders receive shares of new company "B".
(2) Shareholders still own shares of company "A", which now only represent ownership of "A" without "B".

Figure 2.5   Spin-off schematic—The spin-off divestiture at a glance.

Spin-offs are the low-hanging fruit of the investment world. In a spin-off, the parent corporation distributes on a pro-rata basis all the shares it owns in the unit (subsidiary or division) to its own shareholders, thus creating a new public company. This divestiture is typically undertaken for the reasons shown in Table 2.9.

Recipients of the new company's equity usually greet their arrival with great expectations because the newly independent management—freed of corporate yokes—often thrives and generally outperforms the Standard & Poor's stock indexes in the early years. However, on many occasions, stockholders dump their new shares, unwilling to hold a position in a smaller company. Institutions may be forced to sell because their charters forbid owning new issues or small businesses.[13]

2.9.3 The Equity Carve-Out

"Everyone is entitled to be wrong. At least once."

An **equity carve-out (ECO),** also known as Financial Engineering, is a variation of a spin-off. Many companies have chosen to spin off a single subsidiary by means of an equity carve-out; others go further and use the

Table 2.9   Reasons for Spin-Outs

"Business has only two basic functions: marketing and innovation. Marketing and innovation produce results. All the rest are costs." Peter F. Drucker	
• Unit spun-off no longer had "strategic fit"	• Reduction of corporate asymmetries
• Parent's wish to return to its core business model	• Greater customer focus
• Unit creates "pure play" in the market	• Promote innovation
• Unit unprofitable	• Retain and motivate brightest talent
• New management compensation directly tied to unit's performance	• Attract new blood and ideas
• Improved management focus	• Reduce decision-making time
	• Faster introduction of new products

carve-out as a basic organizing principle, repeatedly selling stakes in business units. The private equity industry grew up around corporate carve-outs. Buyout firms believe the strategy, involving the purchase of unloved divisions of big corporations, provides a clear way to add value through strategic and operational improvements.[14]

An ECO, split-off IPO, or a partial spin-off is a form of corporate reorganization in which a company creates a new subsidiary and IPOs it later while retaining control. Usually up to 20% of subsidiary shares are offered to the public. The transaction creates two separate legal entities— parent company and daughter company— each with its own boards, management team, financial, and CEO. Equity carve-outs increase the daughter company's access to capital markets, enabling the new subsidiary to exploit stronger growth opportunities while avoiding the negative signaling associated with a seasoned offering (SEO) of the parent equity.[15] Table 2.10 summarizes the most common types of restructuring associated with a carve-out.[16]

The overarching purpose of a corporate center is to do for the subsidiaries what they cannot do effectively for themselves. Many structures serve this purpose: operating companies, multi-business companies, holding companies, conglomerates, and even investment firms such as Berkshire

Table 2.10   Carve-Out Methods

• First stage of a broader divestiture, preceding:
• Sale of subsidiary to a third party
• Spinoff of remaining ownership to shareholders
• Total or partial company splits
• Tracking stock by creation of new stock class based on divisional valuation

Hathaway. All are different ways for a centralized parent organization to deliver value to its individual business units. The newcomer to the list is the equity carve-out. Like its predecessors, the carve-out enables a subsidiary to draw on the wisdom, experience, and practical assistance of the executive center while offering something new—a degree of independence that appears to foster innovation and growth.[17]

Equity carve-outs, that is, IPOs of subsidiaries, are not unusual. AT&T is the poster child for equity carve-outs, having tried almost every restructuring method known in this galaxy. AT&T underwent a government-mandated carve-out of each "Baby Bell," whereby each shareholder received 1 share of the new company for each 10 A&T shares.

Pharmacia carved out 14% of Monsanto in 2000. In late 1998, DuPont raised $4.4 billion by undertaking an equity carve-out equal to 30% of its oil subsidiary Conoco. Later the same year, CBS raised $3 billion in a carve-out of 16% of subsidiary Infinity Broadcasting. In August 1998, Cincinnati Bell sold 15 million shares in an IPO of subsidiary Convergys. Later that year, Cincinnati Bell distributed the remaining 137 million Convergys shares to its shareholders.

Equity carve-outs offer several advantages to a company and its shareholders. In general, management of the parent firm believes the market value of the separated companies will be greater than the market value of the combined firm prior to the carve-out. Perhaps the investment community has been overlooking the real value of the subsidiary that produces good financial results but is overshadowed by the other parts of the firm. Another plus is that a separately traded stock allows the former subsidiary to use its own stock as a currency for acquisitions and management incentives. The new publicly traded company will have access to the equity markets that can provide capital for expansion.[18]

2.10 Intrapreneurship in Academia

All forms of academic research are inherently intrapreneurial. Many world-class universities such as MIT, Stanford, and Harvard, have technology transfer offices designed to commercialize worthy academic research. Unknowingly, the professors who worked on the grants and developed the technology are intrapreneurs.

Technology Transfer, or Transfer of Technology and Technology Commercialization, is the process of transferring skills, knowledge, technologies, methods of manufacturing, and samples of manufacturing and facilities among universities and other institutions. Such transfer ensures that scientific and technological developments are accessible to a wider range of users who can then further develop and exploit the technology into new products, processes, applications, materials, or services.[19]

The process of commercially exploiting research varies widely, including licensing agreements and establishing joint ventures and partnerships to share both the risks and rewards of bringing new technologies to market. Other corporate actions, for example, spin-outs, are used where the host organization does not have the necessary will, resources, or skills to develop a new technology. Often these approaches are associated with raising venture capital (VC) to fund the development process; VC is more common in the U.S. than in the European Union, which has a more conservative approach to VC funding.[20] Research spin-off companies are popular vehicles of commercialization in Canada, where the rate of licensing of Canadian universities' research remains far below that of the U.S.[21]

There has been a marked increase in technology transfer intermediaries specialized in their fields since 1980, stimulated in large part by the Bayh-Dole Act (and equivalent legislation in other countries), which provided additional incentives for research exploitation. The U.S. Bayh-Dole Act of 1980 allows universities and non-profit institutions, under specific circumstances, to retain intellectual property rights to discoveries resulting from federally funded research (such as SBIR grants). The Act has been credited with stimulating interest in technology transfer activities while generating increased educational opportunities and commercial development.

For example, MIT "intrapreneurs" have started an average of 20 companies per year, based on licenses to MIT technology. Successful start-ups include A123, Akamai, Alnylam, Brontes, E Ink, Ember, Luminus Devices, Momenta Pharmaceuticals, OmniGuide, QD Vision, Xtalic, and Z

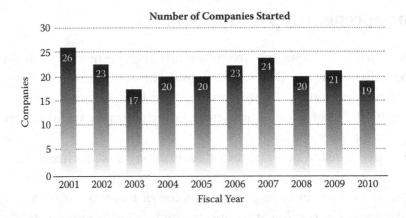

Figure 2.6 Number of companies started—The MIT "intrapreneurs" since 2001.

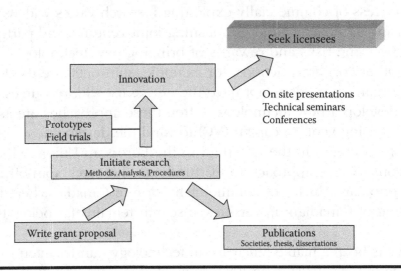

Figure 2.7 Intrapreneurship in academia—Most academic intrapreneurial activity leads to licensing.

Corporation.[22] Figure 2.6 presents the MIT new company startup statistics since 2001.

A typical university-based technology transfer process is shown in Figure 2.7.

2.11 The Ten Commandments of Intrapreneurship[23, 24]

1. Build your team—intrapreneuring is not a solo activity.
2. Share credit widely.

3. Ask for advice before you ask for resources.
4. Under-promise and over-deliver—publicity triggers the corporate immune system.
5. Do any job needed to make your dream work, regardless of your job description.
6. Remember that it is easier to ask for forgiveness than for permission.
7. Keep the best interests of the company and its customers in mind, especially when you have to bend the rules or circumvent the bureaucracy.
8. Come to work each day willing to be fired.
9. Be true to your goals, but be realistic about how to achieve them.
10. Honor and educate your sponsors.

References

1. Pinchot, Gifford & Pinchot, Elizabeth (Fall 1978). Intra-corporate entrepreneurship, Tarrytown School for Entrepreneurs. http://www.intrapreneur.com/MainPages/History/IntraCorp.html, accessed March 2013.
2. McCrae, Norman (April 17, 1982). Intrapreneurial now. *The Economist.*
3. Pinchot, Gifford & Pinchot, Elizabeth (1994). *The End of Bureaucracy & the Rise of the Intelligent Organization.* San Francisco, CA: Berrett-Koehler Publishers, Inc.
4. Investopedia is a premiere resource for investing education, personal finance, market analysis, and free trading simulators. www.investopedia.com
5. http://en.wikipedia.org/wiki/Kelly_Johnson_(engineer)
6. 3M A Century of Innovation.
7. www.wikipedia.org/wiki/Post-it_note
8. Hinskey, Daven (Nov. 9, 2011). www.todayifoundout.com/index.php/2011/11/post-it-notes-were-invented-by-accident, accessed March 27, 2013.
9. Burns, P. (2008). *Corporate Entrepreneurship. Building an Entrepreneurial Organization*, 2nd ed. Houndmills, Basingstoke, Hampshire: Palgrave Macmillan, Chapter 8.
10. *Forbes*, October 9, 1995.
11. *Source*: Thermo Electron SEC filings.
12. Modified after Thompson, T.A. (2007). Sell offs, spin offs, carve outs and tracking stock. http://www.kellogg.northwestern.edu/faculty/thompsnt/htm/d48/ftp/divest_2001.ppt.
13. Luxenberg S. (April 1999). Prince of Spin-offs. *Individual Investor.*
14. Lewis, Toby (Apr. 19, 2010). Corporate carve-outs return to the table. *Financial News*. http://www.efinancialnews.com/story/2010-04-19/corporate-carve-outs-return-to-the-table.
15. http://en.wikipedia.org/wiki/Equity_carve-out
16. Modified after Corporate Restructuring and Divestitures, chapter 11, http://myweb.clemson.edu/~maloney/855/ch11.ppt.

17. http://www.mckinseyquarterly.com/ Equity_carve-outs_A_new_spin_on_the_corporate_structure_203.
18. http://business.yourdictionary.com/equity-carve-out. The American Heritage® Dictionary of Business Terms Copyright © 2010 by Houghton Mifflin Harcourt Publishing Company. Published by Houghton Mifflin Harcourt Publishing Company.
19. http://en.wikipedia.org/wiki/Technology_transfer
20. EU Report on EU/global comparisons in the commercialization of new technologies.
21. State of the Nation 2008—Canada's Science, Technology, and Innovation System.
22. An MIT Inventor's Guide to Startups: for Faculty and Students. http://web.mit. edu/tlo/www/downloads/pdf/Startup_Guide.pdf.
23. Casnocha, Ben (Nov. 28, 2009). The Intrapreneur's 10 Commandments. http:// casnocha.com/2009/11/the-intrapreneurs-10-commandments.html.
24. Pinchot, Gifford. Intrapreneur's 10 Commandments. http:// motivationforultimatesuccess.blogspot.com/2009/07/intrapreneurs-10-com- mandments-by.html.

Chapter 3

Women and Entrepreneurship

3.1 Introduction

As of 2013, it is estimated that there are over 8.6 million female-owned businesses in the U.S., generating over $1.3 trillion in revenues and employing nearly 7.8 million people. Between 1997 and 2013, while the number of businesses in the U.S. increased by 41%, the number of women-owned firms increased by 59%—a rate 1.5 times the national average.[1]

Studies show that entrepreneurial activity by women varies widely across the world. However, women show nearly double the rate of early-stage entrepreneurship in middle-income countries than in high-income countries. These women are highly motivated by opportunity and necessity in middle-income countries.[2]

While male and female founder-entrepreneurs share some basic characteristics, there are some major differences,[3] as shown in Table 3.1.

3.2 Global Impact of Women Entrepreneurs

"You've come a long way, baby!"—Advertisement

According to a report[4] by the Organization for Economic Co-operation and Development (OECD),[5] the number of female entrepreneurs as a share of total was reported as seen in Figure 3.1.

Table 3.1

Characteristics	Female	Male
Level of formal education	High	High
Desire to start own business	High	High
Entrepreneurial experience	Medium	High
Access to capital	Low	High
Mentored by a known industry guru	Low	High
Hindered by family obligations	High	Low
Consider prior experience important	High	Medium
Feel pressure to be family breadwinner	Low	High
Co-founder important to success	High	Medium
Desire to build wealth	High	High
Working for themselves	High	High
Encouragement by peers	High	Low
Having an entrepreneurial friend as a role model	High	Medium

The same report focuses on women and technology because science and technology (S&T) background is a key resource in today's knowledge-based economies. The report concludes,

> "Science and technology gender gaps relate more to attitudes than to aptitudes. Boys are more likely than girls to choose science subjects to study in higher education…. Girls are much less likely to major in computer science, engineering or physical sciences. Although women receive more than half of university degrees in the OECD countries, women account for 30% of degrees in S&T."

The 2nd OECD conference[6] presented data comparing the relationship between self-employed women and men in 28 OECD economies.

We can assume that many of the self-employed women in Figure 3.3 qualify as entrepreneurs under our standard definition. This provides us with a feeling for the substantial economic impact that female entrepreneurs are having throughout the world.

3.3 Personal Challenges Faced by Female Entrepreneurs

"Start small and dream big." —Rich Dad, Poor Dad

By its very nature, entrepreneurship poses significant personal challenges, including money, time, and family pressure. Those challengers can be particularly acute for women, especially those with young children. Moreover, with women starting their own businesses at an increasing rate, understanding the challenges is an important component of success.

Currently female entrepreneurs own almost half of all small businesses in the U.S. The steady rise in female entrepreneurs can be attributed to several factors, most of which share the same rationale as their male counterparts, such as:

■ passion for their ideas
■ desire for economic independence
■ dissatisfaction with current job
■ need to engage in philanthropic causes
■ social imperatives

3.4 Women and Success

"I find that when you have a real interest in life and a curious life, that sleep is not the most important thing." —Martha Stewart

In her seminal 1974 book *Women & Success: The Anatomy of Achievement*, Ruth B. Kundsin[7] asks:

If a female Einstein existed in the United States today, would she be recognized? Would she be at the Princeton Institute for Advanced Studies? Would she get a National Science Foundation grant? Would she be listed in *Who's Who*? Or would she be found in a neat suburban house washing her husband's socks, practicing Craig Claiborne's recipes and imbibing dry martinis in the afternoon with anger mounting in her heart toward her family, her friends and the faculty at the college where she majored in physics?

Another challenge is that women's success hinges on the fact that they are asked to demonstrate competence in a variety of roles, both as women as well as in their occupations. Occupational excellence is insufficient. Men, on the other hand, succeed primarily based on occupational and financial competence.

3.4.1 Defining Success

"Obstacles always precede opportunities."

In the context of this Guide, success will be defined as the unhindered ability to function in a chosen profession with some measure of peer recognition. According to this definition, by the mid-2010s women have indeed achieved success given that female entrepreneurs own almost half of all businesses in the U.S., contribute to the overall employment of 18 million workers, and generate anywhere from $2 to $3 trillion in U.S. economy revenues.[8]

3.4.2 Twenty-Five Leading Female CEOs

According to Catalyst, a non-profit organization, female CEOs are running *Fortune 500* companies, including IBM and PepsiCo. The number has tripled since 2002–2003. The 2012 *Fortune's* annual ranking of America's leading businesswomen lists the following:

Rank 6— > ▼	Name	Company
1	Ginni Rometty	IBM
2	Indra Nooyi	PepsiCo
3	Meg Whitman	Hewlett-Packard
4	Irene Rosenfeld	Kraft Foods
5	Ellen Kullman	DuPont
6	Patricia Woertz	ADM
7	Ursula Burns	Xerox
8	Sheryl Sandberg	Facebook
9	Abigail Johnson	Fidelity Investments
10	Safra Catz	Oracle
11	Melanie Healey	P&G
12	Anne Sweeney	Disney/ABC Television Group

13	Rosalind Brewer	Wal-Mart Stores
14	Marissa Mayer	Yahoo
15	Carol Meyrowitz	TJX Cos.
16	Phebe Novakovic	General Dynamics
17	Sheri McCoy	Avon Products
18	Susan Wojcicki	Google
19	Marillyn Hewson	Lockheed Martin
20	Denise Morrison	Campbell Soup Co.
21	Gisel Ruiz	Wal-Mart Stores
22	Carrie Tolstedt	Wells Fargo
23	Bridget Van Kralingen	IBM
24	Mary Callahan Erdoes	JP Morgan Chase
25	Jan Fields	McDonald's USA

3.5 Hurdles to Overcome

"You'll miss 100% of the bat swings you don't take."

Although women are launching businesses at record pace, they still have an upward climb. Only 29% of the small business loan applicants on Biz2Credit's online platform are from women. Research has found that small business loan approval rates for female-owned companies are 15 to 20% lower than they are for male-owned companies.[9]

According to Biz2Credit, there are three possible reasons why the loan approval rates for female-owned business are lagging:

1. **Female-owned companies are often newer businesses.** They have not established the same track record of success simply because they have not been operating as long as male-owned firms have.
2. **Many female-owned businesses are in industries that banks are wary of funding**, such as nail salons, beauty parlors, and retail franchises. Bankers are not as familiar with the operation of some of these companies, and the industries are perceived as more risky than others are.

3. **Professional services businesses—attorneys, CPAs, doctors, dentists, etc.—tend to be dominated by men.** When these firms request expansion loans or business lines of credit, they encounter far less difficulty in securing capital.

3.6 Role of Angel Groups

It is well accepted that early-stage equity financing in the entrepreneurial sphere plays a critical role. There are two major sources of equity financing for business ventures: (1) business angels in the earlier stages and (2) the institutional venture capital market, which invests primarily in the later stages of the business cycle.[10,11]

Taken together, angels and venture capitalists provide the majority of high-risk equity capital for the startup venture. While women control a substantial portion of the U.S., net worth is a necessary, but not sufficient, condition for attracting angel investments. Angels are typically cashed-out entrepreneurs; individuals who have successfully started and exited their ventures. This pool of female-owned businesses represents the potential female angels of the future.[12] But this female-based source of angel investment is frequently denied to many female entrepreneurs because women investors tend to be more conservative in their investment choices.[13]

Angels rely on their individual network for deal flow, co-investment opportunities, and general guidance in making, monitoring, and due-diligence activities of angel investments.[14] Additionally, this network often serves to spread awareness to other angel investors. If women are outside this traditional network, it often proves to be a significant impediment to participation in angel rounds.

Currently there are well over 150 angel groups operating within the U.S, but less than 25 are affiliated with women. An angel group is an organization that provides a structure and systematic approach for bringing together entrepreneurs seeking early stage capital and business angels searching for investment opportunities. In this context, an angel group is a formal collection of angel investors that meet regularly to examine potential investment deals. The primary goal is to increase deal flow for both parties. As such, angel groups behave closer to professional venture capital firms than to the lone angel of yesteryear.

3.6.1 Angel Capital Market and Female Entrepreneurs

Angel capital is either a first step in eventually qualifying for venture capital or a viable substitute for early-stage venture capital. However, research provides evidence of discrepancies in access to participation between men and women in the angel capital market. Based on annual data from angel organizations between 2000 and 2004, Becker-Blease and Sohl found evidence that female entrepreneurs receive a small portion of the total angel capital awarded.[15] The authors contend that the low rate with which women entrepreneurs are awarded investment dollars in the venture capital market is reflected in the early-stage angel investment market.

3.7 The Green Alliance. A Female-Owned Business

The Green Alliance was born from a partnership of business ingenuity and community environmental advocacy. In 2008, Green Alliance founding business, Simply Green Biofuels, was trying to break into a market dominated by fossil fuels, when it realized that an alliance of similarly green-minded businesses might bring results that are more tangible. Simply Green owner, Andrew Kellar, knew that Seacoast residents using biofuel would similarly be interested in green landscaping companies, green restaurants, etc. Kellar had the idea of teaming up with these other local green businesses, where each business would offer each other's customers mutual discounts, while sharing best business practices and marketing techniques.

At the same time, community activist Sarah Brown was working on community environmental issues in Portsmouth, New Hampshire and Kittery, Maine. As the environmental advocacy widened to include towns and local business, Brown recognized that some businesses, including Simply Green, were leading the way in sustainability. She began to ask how she could bring attention to those local businesses doing the right thing for our planet, as well as encouraging and providing support for the more timid businesses trying to "go green." Hence, the Green Alliance was born.

3.7.1 Profile of Sarah Brown, Owner, Director

Brown's background was far more robust in community activism than entrepreneurship. Brown has been an environmental advocate in her community for the last 10 years. She has served on Kittery's Town Council and the Zoning Board of Appeals. Brown created Kittery Progressive Action, a community action coalition that responds to and organizes around progressive issues in Southern Maine and the New Hampshire Seacoast. She also formed St. John's Stewardship of the Earth Committee in Portsmouth, a group that educates the Seacoast community on environmental issues, and is dedicated to reducing energy consumption and promoting conservation.

Brown spearheaded Cool Kittery, which convinced the Town Council to sign the Mayor's Agreement on Climate Change and then regrouped to become Kittery's current Energy Efficiency Committee—a town-sanctioned board that is now working to reduce Kittery's energy use and promote conservation in the schools and the community. Sarah Brown always cared about the environment and was attracted to initiatives that sought to create a more environmentally friendly community. However, she never dreamed she would go from being a stay-at-home mom of three young girls who, in 2008, served on the Kittery Town Council and Zoning Board, to become the founder and director of the Green Alliance, a Portsmouth-based organization that partners with businesses and consumers to make eco-friendly choices.

Her educational background and initial professional experience in journalism did provide skills that would be put to use in her nascent business. Brown worked for 5 years as a journalist in Moscow, Russia for CNN, BBC, NBC, and the Associated Press. Prior to her experience in Russia, she worked at CNN's New York bureau. Brown earned a degree in Russian studies from Columbia University. She lives in Kittery with her husband and three young daughters.

3.7.2 The Green Alliance Business Model

From the Green Alliance's new office in the Franklin Block on Congress Street, Brown said they now have 93 businesses and nearly 2500 consumer members who are Green Card holders that stretch from Portland, Maine, to Concord and down to Newburyport, Massachusetts. Those businesses include restaurants, dentists, builders, landscapers, printing companies, and merchants.

Brown can often be heard on The River, 92.5 FM, a Haverhill, Massachusetts, radio station, and WSCA Community Radio in Portsmouth giving helpful ecologically friendly tips to listeners on how they can make their world a little greener and better for themselves, their children, and future generations.

By getting businesses to adopt greener practices to reduce waste and use less fossil fuels and getting consumers to patronize businesses that are going green, Brown believes the Green Alliance has made some significant inroads that amount to real cultural change. "We're getting them to think about things they hadn't thought about before," Brown said. "More and more, people are thinking about where they work, where they go out to eat, what they buy for clothes, furniture and how they heat their homes," she said.

Besides recruiting Seacoast businesses and consumers to be "greener," Brown also attends weekend events to promote the Green Alliance's message. She also employs three full-time writers who generate insightful articles for the Green Alliance's websites to show readers how different Seacoast businesses are putting environmentally friendly values into practice.

3.7.3 Business Conceptualization

However, Brown is very quick not to take all the credit for the Green Alliance. She said it was really the brainchild of Andrew Kellar, the owner of Simply Green Biofuels. In 2008, Brown said she and Kellar became friends and Kellar conceptualized the idea and business model for the Green Alliance. Brown formed a partnership with Simply Green Biofuels and Purely Organic Lawncare of York Harbor, Maine.

While Brown was pushing Kittery town officials to create a Green Committee, she discovered that there were some individual businesses like the Beach Pea Baking Co. and Robert's Maine Grill in Kittery that were using green practices. However, they did not have the time or resources to get that message out to the public. "So that was sort of the spark for the idea," she recalled.

Brown actually went to work for Kellar and used her background in journalism to advocate for Simply Green Biofuels. She said Kellar saw the benefit of creating a group that could unite all the businesses that had adopted green business practices. "There was a green customer base that we all could be sharing," Brown said.

With 115 business members, Brown said the region's green sector has more than 1800 employees. The more the Green Alliance can get members of the public to support green businesses, the more those businesses will grow, the more jobs they will create, and the greater the cultural change will be to go greener for more people.

The Green Alliance's Green Card members also enjoy special discounts when they purchase merchandise or services from green businesses who are alliance members. "The consumers, they are the ones who are driving this," she said.

3.7.4 Green Alliance Mission

The Green Alliance mission is composed of four principles:

1. To increase the profits of those businesses having the least negative impact on the environment.
2. To encourage more sustainable business practices through "business-to-business" mentoring and strength in partnership.
3. To educate and influence the public to consider the goods and services they use in their own communities.
4. To encourage more sustainable choices.

3.7.5 The "Green" Business Enterprise

The Green Alliance is one of the "Green Business" or "Sustainable Business" enterprises that have appeared in the American business lexicon in the past two decades. A green business or sustainable business is an enterprise that strives to have minimal negative impact on the global or local environment, community, society, or economy. Often, sustainable businesses have progressive environmental and human rights policies. In general, a business is described as green if it matches the following four criteria:

1. It incorporates principles of sustainability into each of its business decisions.
2. It supplies environmentally friendly products or services that replace demand for non-green products and/or services.
3. It is greener than traditional competition.
4. It has made an enduring commitment to environmental principles in its business operations.

A major initiative of sustainable businesses is to eliminate or decrease the environmental harm caused by the production and consumption of their goods.[9] The impact of such human activities in terms of the amount of greenhouse gases produced can be measured in units of carbon dioxide and is referred to as *carbon footprint*. The carbon footprint concept is derived from ecological footprint analysis, which examines the ecological capacity required to support the consumption of products.[10]

3.8 An Interview with Sarah Brown

Q. In an nutshell, what does your company seek to accomplish?

A. We cover all green sectors including renewable energy, energy efficiency, smart grid, energy storage, green building, the recycling industry, water, organic foods, and healthy lifestyles. We also cover multinational corporate green pioneers that are making a commitment to sustainability and acting on it. But the broader mission is to create a vibrant community of engaged citizens, and the businesses that serve those citizens.

Q. What were your challenges of starting/running a business with small kids at home?

A. When I started, kids were 4, 7, and 8 years old. That was the hardest part. There is a cultural expectation that women will take care of the kids. And it isn't just the logistical problems of working nights and weekends. More debilitating is the extra stress and brutal anguish that affects women. As a mother, you cannot overcome the guilt about the time you're not spending with your family because of your startup. And you take the failures that come with business more personally.

Q. What were the benefits of running the business but reporting to a co-owner with more equity in the company?

A. I had very supportive partners. Being a sole proprietor, I actually wanted a partner to share my emotional burden. Women are used to working collaboratively, and having someone else to share the stress was helpful. It is tough to have the entire burden of running a business on your shoulders.

And while there were certainly times where I wished I had 100% equity, the fact that my business partners were successful businessmen themselves was actually reassuring. More recently, I've had some investor interest. Ideally, if I had a strategic partner or investor, I'd want to the partner to be involved in the business, but I understand that's not always possible.

Q. How did you overcome doubts that a woman with no previous experience running a company could succeed in a difficult economy?

A. I started this enterprise when my partner, who himself was a respected young entrepreneur, said it could be a viable business. And I'm like most other women; outside approval is important to me. That gave me the confidence to move ahead. I was doing activism before starting this business, and that's obviously an unusual path to business ownership. But my activism on environmental and community issues tied into the business concept of connecting citizens and local businesses, as I had a well-developed network of consumers and businesses to approach when building my customer list and growing my sales.

And while the sluggish economy hasn't helped my business, I do stress to my clients that the type of communications, outreach, and lead generation I do—to a well-educated consumer base that will demonstrate above-average brand loyalty—is really important when marketing budgets are lean.

Q. How do you price your services?

A. I charge based on the number of full-time employees a customer has, with higher prices for larger companies, as they have more resources. I've considered basing it on a company's revenues, which would actually be a better way to gauge how much to charge, but then I would be asking for potentially sensitive information from private business owners.

Q. Do women have special business insights?

A. Women are good at recognizing their own weaknesses, and we are also better at admitting weaknesses or mistakes than men. I think that's a positive trait in business; for me it means I've outsourced tasks that I personally don't like to do, so that I can spend the majority of time on the aspects of the business I enjoy, which keeps me motivated.

3.8.1 The Importance of Emotional Support

A supportive spouse or partner can make a significant difference. Brown credits her husband for not only supporting her entrepreneurial spirit, but also picking up much of the slack at dinnertimes, after-school activities, and parental involvement, despite his having a full-time job. "I would not have been able to hold on this long without my husband," Brown declared forcefully. "I'd recommend anyone in a relationship having an honest conversation with their partner about not just the time involved, but the emotional burden you'll both be going through, because that stress of running a business is very hard to keep inside."

The need for support can extend beyond immediate family. Having a co-owner or strategic partner to assist in a new venture can be an immense help. "I was lucky to have supportive partners, and to be getting advice from successful entrepreneurs," Brown said. And a woman's natural tendency to work collaboratively can help mitigate the stress of running a startup.

The business got its start when the co-owner of Brown's employer felt her marketing and business development concept could be a viable business. "Hearing that my idea had support from other successful businesspeople gave me the confidence I needed, since I never even saw myself as a businesswoman."

3.8.2 *Don't Undersell Yourself*

Brown's background is illustrative of the recent trend of more women starting business who weren't born entrepreneurs; her background was in journalism (having served as a foreign correspondent for CNN during the 1990s) and community activism. "I saw how my communication and media skills could tie into my employer's business model, generate leads, and capture additional market share. She also had to learn from other business owners. One client insisted she was working too hard for too little money, and was doing too much work on certain projects that simply didn't generate enough revenue.

In addition, Brown warns that the female tendency to want to help people can work against women entrepreneurs. "I lent professional support to numerous community groups and other entrepreneurs, and I'm proud of the impact I've had on many local businesses. But that was part of the reason that even though I had a great client list, I was still barely profitable."

Brown also cautions against making the mistakes she made when securing her initial client base. "In addition to starting this business at a time when marketing/communications budgets were being slashed, I also underpriced my services. In part, that was because I had pre-existing professional or personal relationships with many of them, and wanted to extend favorable pricing."

Once those precedents had been set, Brown found it hard to raise prices. "Even when I tried to raise my rates modestly, I got a lot of push-back from my clients." She recommends that other women entrepreneurs be careful not to undersell their services when they are selling within their existing network.

Now, Brown has come around on the need for sustainable profitability. While still struggling with profit margins in a tight economy, she wants her company to be successful financially—at least to the point where it matches the emotional support she receives from her community.

3.8.3 Overcoming Perceptions, Gaining Respect, and Finding Success

As for her dealings with clients and others, Brown feels that sometimes she is not respected as much as her male counterparts are. "Some men don't understand that women entrepreneurs go through the same trial by fire." However, she insists that for every "disrespecter," there is a man willing to lend a hand or make an important introduction. Brown's business is up to 115 clients, and continues to enjoy abundant regional respect and influence.

3.8.4 Putting It in Perspective

For women who have done their personal and professional due diligence, and choose to take the plunge into entrepreneurship, Brown's ultimate recommendation is that women should leverage the social and emotional skills they have developed in their careers and personal life to find business opportunities. "There might be opportunities to serve businesses in some way right in your own region that no one's addressing yet." At the same time, she warns women not to let their tendency to be their own worst critics stand in the way of success.

However, despite the slightly higher hurdles, entrepreneurs of both genders share the same excitement about creating and running their own businesses. "I'm not sorry for a minute that I chose be an entrepreneur," Brown concluded. "At this point, I'd probably find it hard to work for someone else!"

References

1. The 2013 State of Women-Owned Businesses Report, commissioned by American Express OPEN. http://www.womenable.com/userfiles/downloads/2013_State_of_Women-Owned_Businesses_Report_FINAL.pdf.
2. Women Entrepreneurship. http://www.authorstream.com/Presentation/rakesh07-1308020-women-entrepreneurship-ppt/.
3. Cohoon, J.M., Wadhwa, V., & Mitchell, L. The Anatomy of an Entrepreneur. Are Successful Women Entrepreneurs Different from Men? Kauffman Foundation of Entrepreneurship, http://www.kauffman.org.
4. Gender and Sustainable Development: Maximising the Economic, Social and Environmental Role of Women. http://www.oecd.org/social/40881538.pdf, Figure 8.
5. OECD, www.oecd.org.

6. http://www.oecd.org/Promoting Entrepreneurship and Innovative SMEs in a Global Economy, Istanbul, Turkey, June 3–5, 2004, p 16, Figure 1.

7. Kundsin, R.B., Ed. *Women & Success: The Anatomy of Achievement*. New York: William Morrow & Co., Inc., 1974.

8. Some facts about women entrepreneurs, Go 4 Funding. http://www.go4funding.com/Articles/Entrepreneur/Some-Facts-About-Women-Entrepreneurs.aspx.

9. Rohit Arora, CEO of Biz2Credit, published March 19, 2013, FOXBusiness.

10. Timmons, J.A., & Sapienza, H.J. Venture capital: the decade ahead, in D.L. Sexton & J.D. Kasants (Eds.), *The State of the Art of Entrepreneurship*, Boston, MA: PWS-Kent, 1992, pp. 402–437.

11. Meyer, R., David, M., Butler, J., Caragannis, E., & Radoevich, R. *The 1995 National Centers of Early Stage Capital Financing*. Albuquerque, NM: Orion Technical Associates, Inc., 1995.

12. Sohl, J.E., & Hill, L. Women business angels: insights from angel groups. *Venture Capital*, 9(3), 207–222, July 2007.

13. Wheelan, D. Investing with care. *American Demographics*, 23(11), 12, 2001.

14. Freear, J., Sohl, J.E., & Wetzel, W.E., Jr. The private investor market for venture capital. *The Financier,* 1(2), 7–15, 1994.

15. Becker-Blease, J.R., & Sohl, J.E. Do women-owned businesses have equal access to angel capital? *Journal of Business Venturing,* 22, 503–521, 2007.

Chapter 4

The Entrepreneurial Environment

4.1 Introduction

"Open the curtains and behind any major innovation and you will find an entrepreneur."

An **entrepreneur** is one who envisions, organizes, directs, funds, and manages a new business. A successful business is one that will build lasting value for the founder as well as for all stakeholders. Figure 4.1 presents, in graphical format, what the entrepreneurial animal looks like.

As the figure shows, many skills have been identified as crucial to entrepreneurs, including initiative, problem solving, perseverance, independent thinking, stamina, commitment, self-confidence, and, above all, an instinct for risk taking—*not gambling, but accepting managing risks*. The entrepreneurial environment can be visualized as shown in Figure 4.2.

Lifelong employment in a large corporation is no longer a preferred objective of many college graduates, as the corporate world is no longer perceived as the ultimate expression of individual success. The "self-made" person has a unique, romantic, and individualistic appeal, particularly if the result is gaining personal wealth while contributing to the nation's overall economic well-being. Becoming financially independent through individual effort—entrepreneurship—is the highest expression of the American dream.

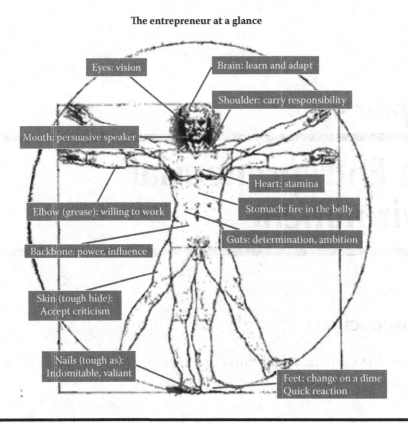

Figure 4.1   The entrepreneur at a glance—Skills crucial to entrepreneurs.

Some individuals appear "entrepreneurial" from a tender age, for example, selling lemonade, cutting grass, or selling cupcakes. Once an inherent entrepreneurial talent has been discovered, most learned entrepreneurs could tip the balance in their favor by following and studying the behavior patterns of experienced entrepreneurs. This chapter will allow you to closely

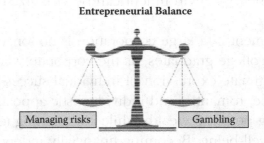

Figure 4.2   Entrepreneurial balance—Successful entrepreneurs are risk managers, not gamblers.

Table 4.1 Murphy's Eight Leadership Roles

1.	Select the right people.
2.	Connect them to the right cause.
3.	Solve problems that arise.
4.	Evaluate progress toward objectives.
5.	Negotiate resolutions to conflicts.
6.	Heal the wounds inflicted by change.
7.	Protect their cultures from the perils of crisis.
8.	Synergize all stakeholders to achieve improvements together.

analyze the secrets behind successful founder/entrepreneurs and to improve your chances of success.

4.2 Leadership Qualities

"Startups are started by leaders."

Leadership can be defined as knowledge and skills that enable a person to use reason, power, and influence to persuade others to follow a desired course of action. Murphy[1] discovered that leadership can be defined and measured as a form of intelligence, exemplified by eight specific roles shown in Table 4.1.

Visually, some of the most crucial leadership skills found in founder/leaders are depicted in Figure 4.3.

4.2.1 Leaders vs. Managers

The major difference between a leader and a manager is behavior patterns. The ability to raise capital, inspire confidence, and nurture the fledgling business is critical to leader-entrepreneurs, not to managers. The differences between leaders and managers are greatly magnified in a startup organization, as noted in Table 4.2.

Thus, according to *our* definition, Lee Iacocca (Ford Mustang), Carly Fiorina (Hewlett-Packard), and Jack Welch (General Electric) are *not*

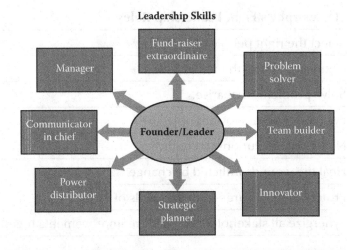

Figure 4.3   Leadership skills—The eight leadership skills crucial to founder/leaders.

entrepreneurs. These individuals did not found new firms with innovations but instead created new wealth for their existing shareholders. In contrast, business leaders such as Bill Gates (Microsoft), Frederick W. Smith (FedEx), Sam Walton (Wal-Mart), Arthur Blank and Bernie Marcus (Home Depot), Howard Schultz (Starbucks), Steve Jobs (Apple Computers), Andrew Carnegie (US Steel), and Michael Dell (Dell Computer) were entrepreneurs.

Table 4.2   Differences between Leaders and Managers in Startups

"Managing is doing things right; leadership is doing the right things." —Warren Bennis and Peter Drucker	
Leaders	*Managers (Traditional Roles)*
Heroic figures	Planning and budgeting
Strategist-in-chief; visionaries	Organizers (coping with complexity)
Leading change and managing chaos	Directors (organizing and staffing)
Problem solvers	Controllers (monitoring activities)
Great negotiators	
Motivators, influencers, innovators	
Corporate spokespersons	
Resource finders	

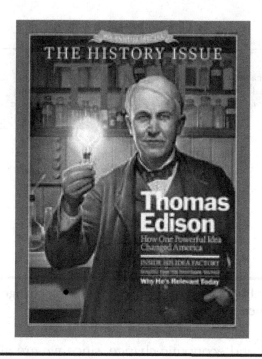

Figure 4.4   Thomas Edison—One of the most prolific entrepreneurs/inventors of all time.

4.2.2 *The Centrality of Leadership in a StartUp*

A startup is a business venture in the early process of developing the infrastructure to operate as a going concern. The entrepreneur-leader (1) foresaw an opportunity, (2) assembled a startup team, (3) performed proof-of-principle studies, and (4) raised enough early-stage capital to initiate operations.

Starting a business "from scratch" takes a lot of guts, hard work, and some luck. However, the leader does not rely on luck; he or she "makes" luck by hard work. Thomas Alva Edison, one of the most prolific of American inventors, was asked to what he attributed his "luck" in inventing so many products. Edison quipped, "The harder I work, the luckier I get." *Time* magazine dedicated an entire issue to Edison, shown in Figure 4.4.

Leadership entails convincing, motivating, inspiring, and, above all, leading and controlling *change*. Veteran entrepreneurs will tell you that the most challenging aspect of their startups was preparing their staff to accept

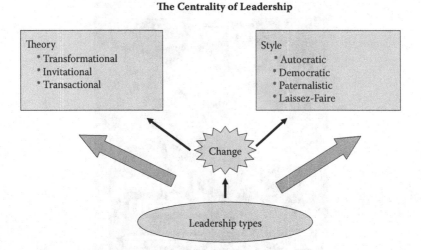

Figure 4.5   **The centrality of leadership—Leaders are the change agents, overcoming organizational resistance.**

fast-paced technical and economic change, since we are all pre-programmed to resist change. This is illustrated in Figure 4.5.

- *Transformational leadership* introduces widespread change to an organization.
- *Invitational leadership* strives to build relationships and a general sense of belonging.
- *Transactional leadership* focuses on management, efficiency, rules, and regulations.

Under an *autocratic style,* the leader makes all decisions without staff consultation or input, whereas the *democratic* leader encourages consensus and majority rule. The *paternalistic style* fosters a "parental figure" who forgives the transgressions of the staff. Finally, the *laissez-faire* (French: "let it be") leader allows decisions to be made by everyone, relying on the ability and goodwill of the entire senior staff.

4.2.3 Leadership Is Situational

"Leadership is a journey. Greatness is its destination."

Leaders find that every senior employee ("follower") requires a different style of leadership, depending on the person's individual development level

Table 4.3   The Four Situational Leadership Styles (in response to follower development level)

S1: Telling/Directing	S2: Selling/Coaching
High task focus, low relationship focus	High task focus, high relationship focus
S3: Participating/Supporting	S4: Delegating/Observing
Low task focus, high relationship focus	Low task focus, low relationship focus
(Follower lead)	(Follower lead)

("maturity"). Thus, leadership is situational. To address this issue, Hersey and Blanchard[2] proposed that leaders adapt their style to a follower development style (or maturity), based on the follower's readiness and willingness to perform required tasks (i.e., competence and motivation).[3] Known as the Situational Leadership Model© (SLM), SLM is most effective when a leader: (1) engages in two-way communication with a follower and asks the follower for input; (2) listens and provides support and encouragement to the follower; (3) involves the follower in decision-making in order to facilitate problem solving; and (4) encourages and promotes self-reliance in the follower's behavior.

Each of four leadership styles (S1 to S4) aligns with a development level (D1 to D4) of a follower. The leadership styles differ in focusing on directing activities vs. emphasizing the leader-follower relationship. This is summarized in Tables 4.3 and 4.4.

Visually, SLM can be summarized in Figure 4.6.[4]

To summarize, the SLM theory states that successful leaders must adapt their leadership style to situational needs. Effective leaders should know that there is not a one-size-fits-all way to manage people, but instead should adapt their styles according to the development levels of the people they lead.

Table 4.4   The Four Development Levels of Followers

D1: Low competence, low commitment/ unable and unwilling or insecure	D2: Some competence, variable commitment/unable but willing or motivated
D3: High competence, variable commitment/able but unwilling or insecure	D4: High competence, high commitment/able and willing or motivated

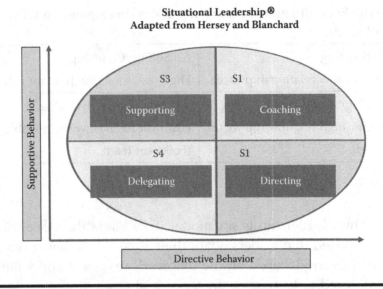

Figure 4.6 Situational leadership—Leaders adapt their style according to the specific situational needs encountered.

4.3 Establishing Your Founder Team

The founder team consists of key personnel and advisers who established the startup organization. Frequently, the team is put together by more than one individual, although typically not all team members have the same amount of initial ownership (founder's shares).

Your business's DNA, including organizational structure, culture, power and influence, and business strategy and tactics, will be influenced in no small part by the characteristics of your team. You will need to decide whether the members of your founding team are *heterogeneous,* that is, diverse, or *homogeneous,* that is, similar to one another, in their abilities and experiences.

The founder with a controlling share majority might ask, "Why do I need to build a great team?" The answer is because lenders and investors prefer an "A" team with a "B" product to a "B" team with an "A" product. Table 4.5 summarizes the most important considerations when building a team.

4.3.1 Elements of Skilled Teams

Skilled teams bring credibility and stability to the startup. The founder will quickly find that his or her area of expertise is not sufficient to ensure business success. Ironically, the founder starts as an expert (Expert = someone

Table 4.5 Characteristics of Founder Teams[5]

• Founder has: (a) higher education credentials, (b) prior successful entrepreneurial experience, (c) recognized expertise in a relevant technical area, (d) professional contacts, and (e) "sweat equity" in the firm.
• A team brings a combination of talents, resources, and experience unmatched by any single individual.
• Frequently, the entire team does not come together at once. Instead, it is built as the new firm can afford to hire additional talent. At first, the firm may rely on non-paid "volunteers" willing the help get the organization "off the ground."
• The team also involves more than insiders. Most startup teams consist of boards of directors, boards of advisers, and other professionals on whom they rely for direction and advice.
• New ventures have a high propensity to fail, in part, due to what researchers call the "liability of newness." Startups often falter because the people who start firms cannot adjust quickly enough to their new roles and the firm lacks a "track record" with investors, buyers, and suppliers.

who knows more and more about less and less) but later must become the great "generalist."

The skilled team brings access to all areas of business expertise, including marketing, sales, manufacturing, accounting, and finance. Startups survive only by standing on the shoulders of many people, especially those with collective experience in a multitude of disciplines. A team provides greater opportunities to network, that is, build and maintain relationships with people whose interests are similar or who can bring strategic advantages to a firm.[6] This is illustrated in Figure 4.7.

4.3.2 Establishing Your Board of Directors

"Quality, not quantity."

Organizing a venture as a corporation requires constituting a board of directors (BOD), a slate of individuals elected by a corporation to oversee the governance of the entity. Corporate governance involves regulatory and market mechanisms, relationships between a company's management and its board, shareholders and other stakeholders, roles and responsibilities of senior individuals, and the goals that the corporation aims to achieve.[7] The best boards do more than govern; they add value to your firm.

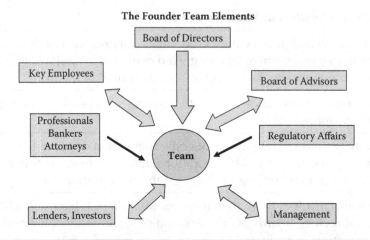

Figure 4.7 The founder team elements—The stakeholders of the new enterprise.

In addition to fulfilling traditional roles of setting and monitoring implementation of business strategy, boards increasingly provide greater contributions in operations, succession planning, executive compensation, risk management, tactical decision-making, marketing strategies, etc.

Startups should start with an odd number of directors (typically three or five) to prevent potential deadlock on contentious issues. Aside from the founder, who usually serves as Board Chairman, there is a tendency to evenly split between inside directors (persons who work for the company) and outside directors (directors who are independent and unaffiliated with the firm) to ensure examination of issues from a balanced perspective.

The greatest benefit of a BOD is its knowledge and understanding of the industry. Good directors closely examine strategic plans, assumptions, aggressiveness, and current market realities. Furthermore, the BOD should collectively bring expertise in specialized areas such as financial auditing, fundraising, regulatory matters, and overseas operations.

The BOD has primary responsibility for overseeing and implementing the company's governance system. While responsibility for performing various functions is delegated to specific departments, the BOD consolidates all the activities under one coherent set of rules, as depicted in Figure 4.8.

4.3.3 Selecting Your Board of Advisors

A Board of Advisors (BOA) is a panel of experts convened by the founder to provide counsel, advice, and direction on an ongoing basis. Importantly,

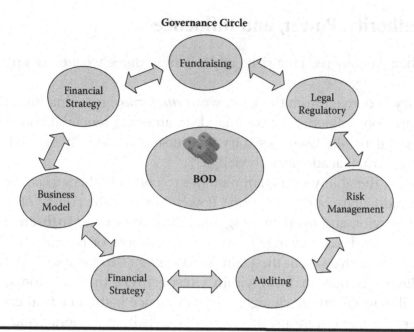

Figure 4.8   Governance circle—The BOD is primarily responsible for the governance of the enterprise.

members of the BOA do not have legal liability to the firm because their opinions are considered advisory and non-binding. Legally and operationally, BOAs are established to provide expert opinions to management on specific issues, and serving on a BOA is less time-consuming than serving on a BOD.

Table 4.6 presents some important considerations when choosing your BOA members.

Table 4.6   Considerations When Choosing BOA Members

Do the candidates possess the technical skills relative to the firm?
Are they well known and respected in their field?
What is their success track record in similar small organizations?
Do the candidates have experience dealing with startups?
Have the candidates published papers in relevant technical journals?
Can the candidates get along with experts in other fields?

4.4 Authority, Power, and Influence

"When we disagree I am right, and when we agree we are both right."

Authority is derived from the Latin word *auctoritas*, meaning invention, advice, opinion, influence, or command. In an entrepreneurial context, *authority* is generally used to signify the persuasive capacity derived from possessing formal academic knowledge.[8]

Power is the ability of a team member to compel others to act or to accept a particular course of action.[9] Power can alter the course of events, change behavior, and overcome organizational resistance. **Influence** is the ability to effect desired change by virtue of authority or power. *Power* and *influence* are derived from the Latin words *potere* ("to be able, capable") and *influere* ("to flow in, from within"), respectively. In other words, power is the ability to effectuate change, while influence is the practical exercise of power.[10] In many startups, the founder exerts both authority *and* power.[11] Initially, the founder is the recognized and uncontested leader by virtue of the expert knowledge that allowed the establishment of the firm in the first place.

Power is derived from five recognized sources: legitimate, reward, coercive, expert, and referent,[12] as depicted in Figure 4.9.

Legitimate power is evidenced by formal titles, for example, CEO, President, or Vice President. The founder starts with a great deal of legitimate power because the dual titles of Chairman of the Board and President are at the top rung of the organization.

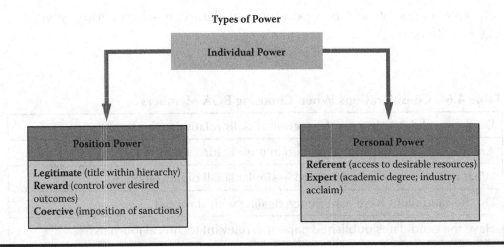

Figure 4.9 Types of power—The entrepreneur's power grid.

Reward and coercive powers are highly influential based on control over distribution of valuable compensation (salaries, awards, recognition) or ability to punish unwanted behavior.

Referent power derives from the possession of or access to scarce resources (e.g., budgets, meeting schedules, attendance at technical conferences).

Expert power is based on advanced technical degrees, specialized skills, information, knowledge, and expertise ("Knowledge is power").

At this point, we need to emphasize a crucial distinction between authority and power. Because authority and status are derived from personal knowledge, such knowledge is perpetually relevant unless made obsolete by new technology. Conversely, power is frequently dependent upon reciprocity ("If you agree to do this for me, I will owe you one"), power is constantly diminishing. Power can be likened to a car battery that can provide a charge when needed, but it needs periodic re-charging, as illustrated in Figure 4.10.

As the saying goes: "Power is the ability of letting others have it your way."

4.5 The Founder and Organizational Politics

"It is not about who is in charge, but who takes charge."

Whether or not the founder wants it, organizational politics occurs in every business environment. In the business environment, politics is the informal struggle for personal dominance where each player struggles to make his or her abilities and achievements known to the group.[13]

Any time people associate, politics will be part of the interaction, and a startup is no exception. In fact, startups are ripe for politics because the

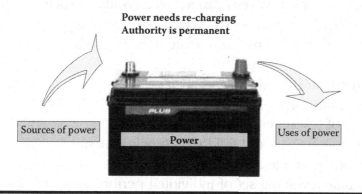

Figure 4.10   Power needs re-charging—Power is dependent on reciprocity ("re-charging").

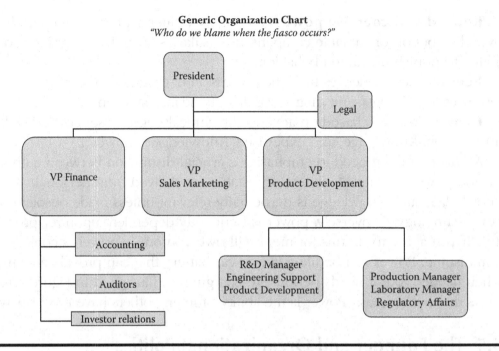

Figure 4.11   Generic organization chart—A typical official organization chart.

official organizational chart may not accurately reflect the real leaders and followers. Figure 4.11 is an "official" (albeit) generic organizational chart.

However, in reality, there exists a phantom, parallel, unofficial "chart" that more accurately reflects those who get "things done," as shown in Figure 4.12.

A parallel organization differs from a traditional, hierarchical organization in that it reflects organizational politics. It is possible—and sometimes advantageous—to blend an official structure with a parallel structure. For example, a subset of employees can act as a parallel structure to develop solutions to specific problems (i.e., situational leadership).

Organizational politics are most acute in a startup setting when there is:

■ high emotional insecurity
■ a chaotic atmosphere
■ a scarcity of resources
■ a highly competitive and fluid work environment
■ an impending financial or technological crisis
■ a continually evolving set of individual performance standards[14]

We will discuss organizational politics more fully in Section 4.5.1.

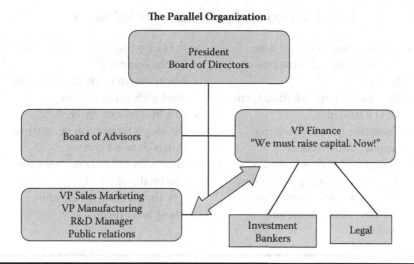

The Parallel Organization

Figure 4.12 The parallel organization—This chart more accurately reflects reality (politics).

4.5.1 Power, Politics and Influence

"Successful companies did not start from third base."

Startups must be built from the ground up, and politically skilled and savvy employees are among the most effective builders of the company. Ahern defined political skill as "the ability to effectively understand others at work, and to use such knowledge to influence others to act in ways that enhance one's personal and/or organizational objectives."[15] We will call these employees "actors."

Like it or not, politics is a fact. Actually, politics is one of the most important mechanisms that enable actors to get things done quickly and efficiently. The politics of the organization even helps to shape culture, and vice versa. Political moves occur so rapidly that managers can get things done with minimum resistance.

There is often a negative connotation when the word politics is whispered. However, politics is far from being a negative; only bad actors give it a bad name. When properly applied, politics is an important managerial tool of the actor. The unpleasant name comes from the few unethical actors who play "dirty politics" mainly for their own benefit, not for the organization as a whole. The best actors are seldom recognized as such by those around them.

Characteristics of politically skilled/savvy

- Actively promote their boss
- Correctly identify true power brokers
- Constantly improve their technical knowledge
- Volunteer to write meeting agendas and minutes
- Design and author business documents dealing with rules, regulations, procedures, etc.
- Are part of the dominant coalition

- Publicize their activities as "best for the company"
- Always seem ready to tackle the most difficult problems
- Are "doers" not "talkers"
- Develop a reputation for getting things done on time
- Their solutions to problems are practical and timely
- Are always there when you "most need them"

Figure 4.13 Characteristics and Performance Attributes of political cognoscenti.

To advance organizational objectives, politically skilled leaders form coalitions to secure their power bases and to protect their occupancy of powerful, central positions in the organization. The most powerful of these coalitions is called *dominant coalitions*,[16] and wields the greatest power in the organization.

Politically skilled/savvy individuals share some important performance characteristics, summarized in Figure 4.13.

The main trick of politics is to make it appear rational and in the best interest of the organization (not the political actor). In addition, politics is best done in the shade. "The rattlesnake that survives is the one that doesn't rattle." Importantly, the founder is above all this maneuvering, possessing authority by virtue of title, knowledge, and voting power. This is depicted in Szycher's Founder's Supremacy Theory, illustrated in Figure 4.14.

4.5.2 Organizational Persona

The founder sets the tone by his or her "imprimatur" on both the political environment and the persona of the organization. In every respect, the firm reflects the personality of its founder, and the dominant coalition moves the

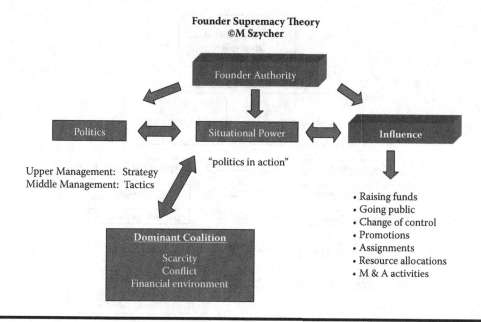

Figure 4.14   Founder Supremacy Theory—The founder's authority trumps all political maneuvering at the start.

company forward via organizational politics. Organizational politics can be defined as "intentional acts of influence to enhance, protect and advance the self-interests of individuals."[17]

The founder decides the organizational persona by his or her initial decisions as presented in Figure 4.15.

Organizational persona
as determined by founder

- Hunting?   Or   Fishing?
- Manufacture? Or   Outsource?
- Gazelle?   Or   Turtle?
- Family?   Or   Harvest?
- Flexibility?   Or   Control?
- Lifestyle entrepreneur (sacrifice income for relaxed work habits) .... or ....
- High value builder entrepreneur

Figure 4.15   Organizational persona—What culture "persona" do you want your organization to adopt?

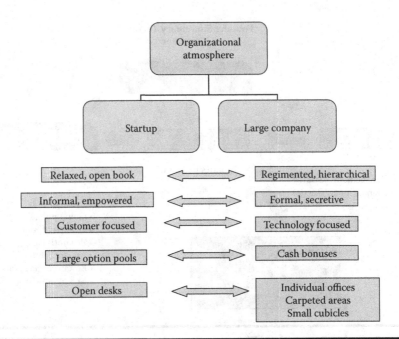

Figure 4.16 Organization management—The founder must decide how to manage the startup.

Last, the founder decides how the organization will be managed, as shown in Figure 4.16.

4.6 The Dominant Coalition

"Chaos creates opportunity."

The dominant coalition is a select group of people within an organization—mostly the executive or senior management team—making all important decisions regarding the direction and focus of the firm. The term "dominant coalition" was coined by James Thompson who adopted the work of March and Cyert[16] in his 1967 book entitled *Organizations in Action*.[18]

In business, coalitions have been present for many years to bring people, departments within an organization, entire companies, or industries together for a common purpose. Examples of such purposes include achieving an important corporate goal, lowering entering a market, regulating an industry action, or strategic planning. A coalition is an exercise in power, whether in politics or business.[19]

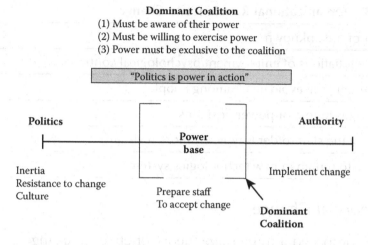

Figure 4.17 Dominant coalition—Visual depiction of how the dominant coalition operates behind the scenes.

Many managers form coalitions to (1) secure their power source, (2) protect their power positions, and (3) wield the maximum amount of power and influence. Dominant coalitions are formed through compromise, bargaining, and negotiating among managers from different functions.[20]

Figure 4.17 summarizes the *modus operandi* of a typical dominant coalition.

4.6.1 Dominant Coalition as Change Agent

"The great pleasure in life is doing what people say you cannot do."
—Walter Bagehot

Any proposed change to the status quo, no matter how desirable, will be fiercely resisted at all levels. General Motors, Sears, JC Penney, Wang Computers, and Polaroid faced the need for dramatic change, but each failed. Managers pushing for organizational or technological improvements were ignored or, worse yet, fired. These companies could not overcome cultural and behavioral obstacles or the power structures inherent in their organizations.

The main reasons for organizational resistance to change are presented in Table 4.7.

Table 4.7 Organizational Resistance to Change

• Fear of the unknown
• Renegotiations of employment/psychological contracts
• Inherent "risk avoidance" among people
• Changes in role or power positions
• Age of the staff (older is typically more resistant)
• Need to retrain in new technologies/systems

4.6.1.1 Theory of Change

Kurt Lewin[21] proposed a three-stage theory of effective change commonly referred to as **Unfreeze, Change, Freeze** (or Refreeze). The dominant coalition could follow the Lewin sequence, as shown in Table 4.8.

Table 4.8 Productive Change Sequence

• Impetus for change: Create dissatisfaction and crisis
• Fear
• Vision of a better tomorrow
• Opportunity for all to advance
• Unfreeze the system
• Reject old customs and approaches
• Reduce support for old ways
• Separate from past practices
• Reward those engaged in new approach
• Change/psychological conversion
• Rejection of the past
• Separation from the past
• Practice the "new" until new system is internalized and understood
• Refreeze the new ways
• Re-enforce new attitudes
• Reward new behavior patterns
• Be consistent, predictable

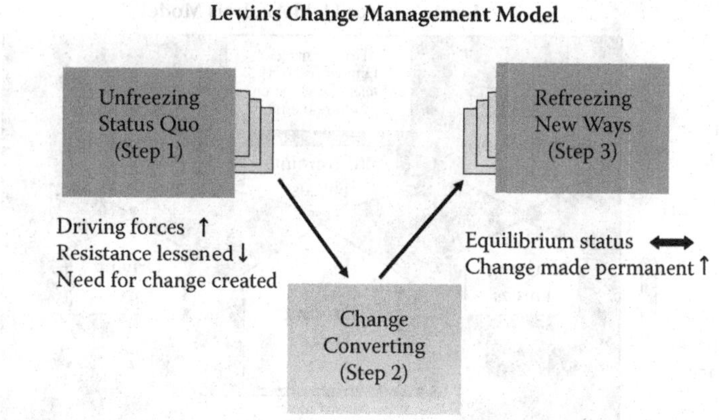

Lewin's Change Management Model

Figure 4.18   Lewin's change management model—To create change, you first "unfreeze" the status quo, and then "refreeze" in the desired new way.

The Lewin three-stage theory of change model can be more easily visualized as shown in Figure 4.18.

4.6.1.2 Force Field Analysis

When first proposed, the force field analysis (FFA) became an influential development in the field of social science. Originally applied to social settings, FFA provided a framework for looking at factors (forces) that influence a situation. FFA looks at forces that either drive toward achieving a goal (helping forces) or block movement toward a goal (hindering forces). The FFA principle as developed by Lewin[22] was a significant contribution to the fields of organizational development, process management, and change management.[23]

The "field" was conceived as very dynamic, changing with both time and experience. When fully constructed, an individual's "field" (Lewin used the term "life space") describes that person's motives, values, needs, moods, goals, anxieties, and ideals.[24] Figure 4.19 presents an idealized conception of FFA.

4.7 Leaders Need Followers

"Leaders rarely use their power wisely or effectively over long periods unless they are supported by followers who have the stature to help them do so." —Ira Chaleff

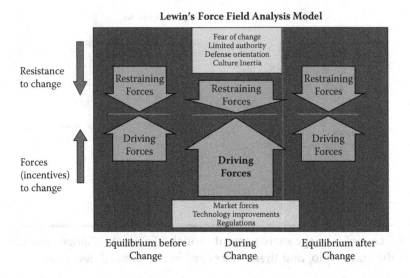

Figure 4.19 Lewin's force field—Visual depiction of opposing forces that shape goal achievement.

Followership is a leader's purposeful guidance and direction in the business environment. Organizations are successful because not only can their leaders lead, but also how decidedly are followers willing to follow. Followership is the reciprocal organizational process of leadership.[25]

Studying followership is integral to better understanding leadership because the success and failure of groups, organizations, and teams is dependent on effective leading and following. Specifically, followers play a crucial and active role in organization, group, and team successes and failures.[26]

The literature on leadership is deep but, not surprisingly, shallow on followership. The emergence of the field of followership in 1988 is widely credited to Kelly in his influential *Harvard Business Review* (HBR) paper "In Praise of Followers."[27] Kelly argued that an effective follower is distinguished from an ineffective one by intelligent, responsible, and enthusiastic participation in pursuing organizational goals. Kelly further described four main qualities of effective followers:[28]

1. **Self-Management:** Ability to think critically, to control one's actions, and to work independently. Followers must manage themselves effectively so leaders can delegate tasks to these individuals.
2. **Commitment:** Commitment to the goals, visions, or causes of a group, team, or organization. This is an important quality of

followers, as it helps keep one's (and other members') morale and energy levels high.

3. **Competence:** Possessing skills and aptitudes necessary to complete the goal or task. Individuals who value this quality often demonstrate more acute skills than their average co-worker (or team member), continually upgrading their skills through classes and seminars.

4. **Courage:** Effective followers hold true to their beliefs and maintain and uphold ethical standards, even in the face of dishonest or corrupt superiors (leaders). These individuals are loyal, honest, and, importantly, candid with their superiors.

4.7.1 Current Views of Followership

Followership has been popularized by Chaleff under the title "The Courageous Follower."[29] According to Chaleff, effective followers need to exhibit three qualities to fully assume responsibilities, as seen in Figure 4.20.

1. **Followership intrinsic power.** Followers have far more power than generally acknowledged. Followership entails accepting and understanding the sources of its power, working towards the common goal, and using available tools to achieve the mission.

2. **Appreciate leader's value.** Understand the pressures placed upon the leader, and work to minimize those pressures to contribute to the common purpose.

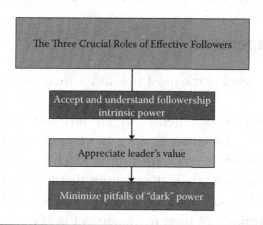

Figure 4.20   The three crucial roles—Leaders must have effective followers.

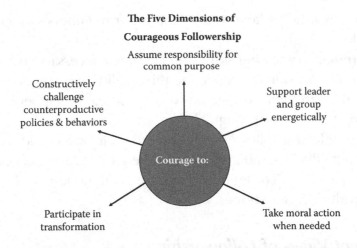

The Five Dimensions of

Courageous Followership

Assume responsibility for
common purpose

Constructively
challenge
counterproductive
policies & behaviors

Support leader
and group
energetically

Courage to:

Participate in
transformation

Take moral action
when needed

Figure 4.21   The five dimensions—Followers must be motivated to act courageously.

3. **Minimize pitfalls of "dark" power** by helping the leader to stay on track. If power becomes corrupt, the follower should speak up and provide constructive feedback to the leader.

Courageous followers share five characteristics as shown in Figure 4.21.

4.7.2 The Followership Universe

It is important to understand that followers and leaders make equivalent contributions to sustaining organizational viability and advancing the organization's goals. Followers within knowledge-based organizations are distinguishable at all organizational levels and activities. Figure 4.22 presents an accurate representation of the interactions between followers and their leader.

4.8 Time Management

"The entrepreneurial clock must tick faster than anyone else's."

Time is your most implacable enemy. We all have the "same" amount of time: 24 hours a day, 7 days a week, etc., but entrepreneurs must break the "tyranny of time." You must accomplish more each day than the "big guys." In 1748, Ben Franklin exclaimed, "Remember that time is money." To utilize limited time efficiently and effectively, you must practice time management.

The standard definition of time management is planning and controlling how time is spent with the goal of effectively accomplishing your

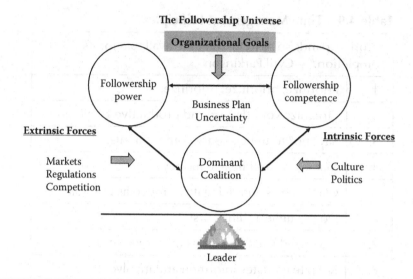

Figure 4.22   The followership universe—Leaders and followers make equivalent contributions toward operational viability.

goals. Poor time management may be related to procrastination or lack of self-control. Managing your time involves planning for the future, setting goals, prioritizing tasks, and monitoring where your time actually goes.[30]

Table 4.9 summarizes some of the important principles behind practical time management.

4.8.1 The Urgent-Important Matrix

A time-honored technique used in business management is categorizing large amounts of data into groups. These groups are often denoted Quadrants 1, 2, 3, and 4, as they appear in Figure 4.23. Each activity is categorized into a quadrant based on these general criteria:[31]

- **Quadrant 1**: Tasks that are perceived as being urgent and important
- **Quadrant 2**: Tasks that are urgent but not important
- **Quadrant 3**: Tasks that are not urgent but important
- **Quadrant 4**: Tasks that are neither urgent nor important

Based on Pareto's law (i.e., the 80/20 rule), 80% of tasks can be completed in 20% of the available time while the remaining 20% of tasks will take up 80% of the time. Following Pareto's law, it is recommended that tasks that fall into the first category (i.e., 80% of them can be completed in

Table 4.9 Time Management Principles

"Work expands so as to fill the time available for its completion." —Cyril Parkinson	
1.	Prioritize, prioritize, prioritize.
2.	Organize your day around productive activities.
3.	Say "no" to unnecessary commitments.
4.	Manage your time like money.
5.	Clarify goals and delegate as much as possible.
6.	Do the ugliest things first.
7.	Make and keep deadlines (no exceptions).
8.	Keep teammates informed and involved.
9.	Touch each document only once.
10.	Keep meetings productive and short.

20% of the time) be assigned a higher priority. If time management aims to maximize productivity, these tasks should receive higher priority. Thus,

Quadrant 1 is reserved for critical activities (decision-making, instructing, commanding, coaching, raising capital)

Quadrant 2 consists of important goals (mentoring, empowering)

Quadrant 3 may contain interruptions (firefighting, crises)

Quadrant 4 activities should be totally disregarded (trivia, reading mail, long phone calls)

Applying the Pareto principle (as described earlier) depends on the method adopted to complete the task; that is, there is always a simpler and more efficient way to complete the task. Employing a complex way will be time-consuming, so the efficient entrepreneur should always try to find out alternate ways to complete each task.[32] This is summarized in Figure 4.23.

4.9 Conflict Management

"Fight as if you are right; listen as if you are wrong."

If you are running a fast-paced organization, you will surely face organizational conflict. Conflict is a natural, recurring, and inevitable outcome of group dynamics. How will you manage conflict?

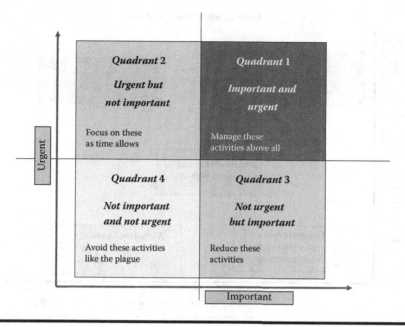

Figure 4.23   Urgent-important matrix—Differences between urgent and important.

Regarding organizational conflict, leaders come in two flavors: (1) the production or task-oriented leader and (2) the people-oriented or accommodative leader. Of course, these are two polar opposites, and many leaders fall somewhere in between. Neither extreme is necessarily right or wrong, but instead will depend on the situation. It is useful for you to understand your inherent leadership characteristics so you can devise a suitable solution when conflicts inevitably arise.

A useful framework for analyzing yourself is the Blake-Mouton Managerial Grid, also known as the Leadership Grid.[33] This model originally identified five distinct leadership styles based on *concern for people* and *concern for production*. The optimal leadership style in this model is based on Theory X and Y.[34] (In Theory Y, management assumes employees are ambitious and *self-motivated,* and typically exercise *self-control.* In Theory X, management assumes employees are inherently lazy and will avoid work if possible because they inherently dislike work. Not surprisingly, Theory X has been proven counter-effective in most modern studies.)

The grid plots your task-centeredness (i.e., concern for production) vs. your people-centeredness (i.e., concern for people). By plotting "concern for production" on the abscissa against "concern for people" on the ordinate, we can surmise that leaders who place too much emphasis in one area at the expense of all others will likely experience lower productivity and

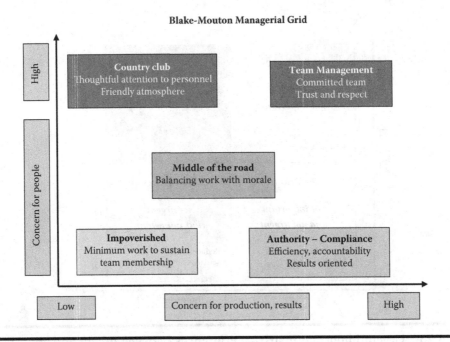

Figure 4.24 Blake-Mouton managerial grid—A useful framework for self-analysis as a leader.

prolongation of conflict. While the grid does not entirely address the question of which leadership style is "best," it provides an excellent starting place for critically analyzing your own performance and improving your leadership skills.[35] This is shown in Figure 4.24.

The Managerial Grid is based on two behavioral dimensions:

- **Concern for People**—The degree to which a leader considers the needs of team members, their interests, and areas of personal development when deciding how best to accomplish a task.
- **Concern for Production**—The degree to which a leader emphasizes concrete objectives, organizational efficiency, and high productivity when deciding how best to accomplish a task.

4.10 Entrepreneurial Rewards

"My idea of risk and reward is for me to get the reward and others to take the risk."

The American dream involves two characters: (1) the entrepreneurial hero and (2) the industrial drone. Former U.S. Secretary of Labor Robert

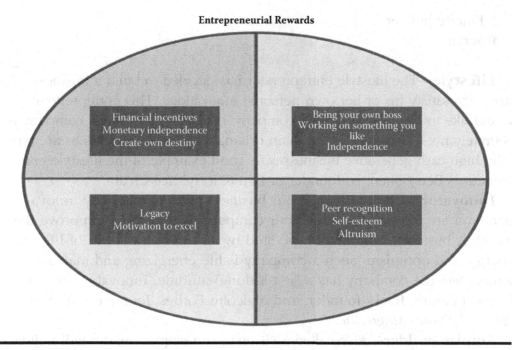

Entrepreneurial Rewards

Financial incentives
Monetary independence
Create own destiny

Being your own boss
Working on something you like
Independence

Legacy
Motivation to excel

Peer recognition
Self-esteem
Altruism

Figure 4.25 Entrepreneurial rewards—More than just money, entrepreneurs are showered with many non-monetary rewards.

Reich described the two as "the inspired and the perspired." These characterizations highlight the difference between the entrepreneur, the seeker of opportunity, and the administrator, the guardian of resources and procedures.

Any entrepreneurial hero will tell you that individual recognition is more important than salaries, bonuses, and promotions. For many people, the dream of charting their own destiny and being self-sufficient is enough to stimulate the pursuit of opportunity and a willingness to invest time and money.

Figure 4.25 summarizes the most compelling rewards listed by entrepreneurs.

4.11 Entrepreneurial Types

"Find your sweet spot."

Not all entrepreneurs are born the same. Pioneering studies by Cole[36] defined four distinctive types of entrepreneurs. More recently, researchers have refined the style categories as:

1. Lifestyle
2. Innovator

3. Empire builder
4. Serial

Lifestyle—The lifestyle entrepreneur has decided to build a business strictly to satisfy his or her own personal motivations. This entrepreneur would like to create a successful company, but building a public company is definitely not a necessity or the main driving force. Such businesses are typically high cash generative businesses. A good example of the lifestyle entrepreneur is Ben Cohen, co-founder of Ben & Jerry's Ice Cream.

Innovator—If you operate your business predominately by innovation, you are focused on using your company as a means to improve the world. A business owned and operated by an innovator is full of life, energy, and optimism. Such a company is life energizing and makes customers feel the company has a "get it done" attitude. Innovators include Gordon Moore, Intel's founder, and Malcolm Forbes, founder and publisher of *Forbes Magazine*.

Empire builder—Also called a visionary, an empire builder will build a business based on his or her future vision. The empire builder requires a high degree of curiosity to understand the surrounding world and devise plans to avoid the landmines. An example of an empire builder is Bill Gates, founder of Microsoft.

Serial—This entrepreneur's main motivation is to build a business with a clear exit strategy (to "harvest" the enterprise). Often this entrepreneurial type is ignored by many angel investors because the serial entrepreneur may have mixed priorities, and may have started numerous business ventures and then moved on! Interestingly, some angel syndicates actually do prefer to invest in the serial entrepreneur because this entrepreneurial type is focused on the exit or sale, and the cash payout is the main motivation.[37] Serial entrepreneurs include Donald Trump, founder and CEO of Trump Hotels & Casino Resorts.

4.12 The Ten Commandments of Leadership

1. Thou shall balance influence with good judgment.
2. Thou shall speak authoritatively and clearly.
3. Thou shall prioritize achievable goals.
4. Thou shall delegate to motivate.
5. Thou shall promote diversity of skills.

6. Thou shall clearly communicate your vision/mission.
7. Thou shall inspire confidence by example.
8. Thou shall focus on results, not activities.
9. Thou shall encourage mentoring at all levels.
10. Thou shall develop and publicize a succession plan.[38]

References

1. Murphy, E.C., *Leadership IQ*, New York: John Wiley & Sons, Inc., 1996.
2. Hersey, P., & Blanchard, K.H. *Management of Organizational Behavior*, 6th ed. Englewood Cliffs, NJ: Prentice Hall, 1993.
3. http://changingminds.org/disciplines/leadership/styles/situational_leadership_hersey_blanchard.htm
4. Situational leadership. http://web.usf.edu/airforce/AS300/SLD19_Situational_Leadership.ppt, accessed April 2013.
5. Modified after Barringer, B.R., & Ireland, R.D. Building a New-Venture Team. http://wps.pearsoned.co.uk/wps/media/objects/8940/9155051/ema_ge_barringer_entrep_3/barringer_e3_ppt_09.ppt. Pearson Education, 2010.
6. Barringer, B.R., & Ireland, R.D. Building a New-Venture Team. Chapter 9. Pearson Education, www.pearsoned.com.
7. http://en.wikipedia.org/wiki/Corporate_governance
8. http://en.wikipedia.org/wiki/Authority
9. Modified after Power and Influence. http://web.usf.edu/airforce/AS300/SLD_26_Power_Influence_08.ppt
10. South Australian Centre for Leaders in Education. Influence and Influencing. Eyre District Leaders. August 2007. http://www.decd.sa.gov.au/eyreandwestern/files/links/PPInfluence_Topic_Aug_07.ppt.
11. Pfeffer J. *Managing with Power*. Boston, MA: Harvard Business School Press, 1992.
12. French, J.P.R. Jr., & Raven, B. The bases of social power. In D. Cartwright and A. Zander (Eds.), *Group Dynamics*. New York: Harper and Row, 1960, pp. 607–623.
13. Gordon, C. Office Politics and Gossiping. http://www.careersmarts.com/shrm/media/office_politics.ppt
14. Modified after Artra, L. Surviving Office Politics. http://www.sasag.org/2004/04/SA_2_OfficePolitics_v2.ppt, 2004.
15. Ahearn, K.K., Ferris, G.R., Hochwarter, W.A., Douglass, C., & Ammeter, A.P. Leader political skill and team performance. *Journal of Management,* 30, 309–327, 2004.
16. Cyert, R.M., & March, J.G. *A Behavioral Theory of the Firm*. Englewood Cliffs, NJ: Prentice-Hall, 1963.
17. Influence, Power, and Politics: An Organizational Survival Kit. Chapter 13, http://www.drluisortiz.com/PPT/ob09Chap013.ppt, McGraw-Hill, 2010.

18. Thompson, James D. *Organizations in Action: Social Science Bases of Administrative Theory.* New York: McGraw-Hill, 1967.
19. http://www.enotes.com/coalition-building-reference/coalition-building.
20. The Power-Control Model. Power of Contingent/Variables. http://www.iun.edu/~bnwcls/w430/orgpol.ppt.
21. Lewin, K. Group decision and social change, *Readings in Social Psychology*, E.E. Maccoby, T.M. Newcomb, and E.L. Hartley (Eds.). New York: Holt, Rinehart and Winston, 1958, pp. 197–211.
22. Lewin, K. Defining the "Field at a Given Time." *Psychological Review,* 50, 292–310, 1943. Republished in *Resolving Social Conflicts & Field Theory in Social Science,* Washington, D.C.: American Psychological Association, 1997.
23. http://en.wikipedia.org/wiki/Force_field_analysis
24. Cartwright, D. Foreword to the 1951 Edition. *Field Theory in Social Science and Selected Theoretical Papers-Kurt Lewin.* Washington, D.C.: American Psychological Association, 1997. Originally published by Harper & Row.
25. Riggio, R.E., Chaleff, I., & Blumen-Lipman, J. *The Art of Followership: How Great Followers Create Great Leaders and Organizations.* San Francisco, CA: Jossey-Bass, 2008.
26. Baker, S.D. Followership: The theoretical foundation of a contemporary construct. *Journal of Leadership & Organizational Studies*, 14, 50–60, 2007. doi:10.1177/0002831207304343.
27. Kelly, R.E. In praise of followers. *Harvard Business Review*, 142–148, Nov-Dec, 1988.
28. http://en.wikipedia.org/wiki/Followership
29. Chaleff, I. The Courageous Follower: A New View of Leader-Follower Relationships. http://www.tobiascenter.iu.edu/conferences/documents/UofINDIANAsharedfile.ppt, 2009.
30. http://en.wikipedia.org/wiki/Time_management
31. Modified after Lakein, A. *How to Get Control of Your Time and Your Life.* New York: P.H. Wyden, 1973.
32. http://en.wikipedia.org/wiki/Time_management
33. Blake, R., & Mouton, J. *The Managerial Grid: The Key to Leadership Excellence.* Houston: Gulf Publishing Co., 1964.
34. McGregor, D. *The Human Side of Enterprise.* New York: McGraw Hill, 1960.
35. Zeida, H. The Blake Mouton managerial grid, *The Certified Accountant*, 39, 82–85, 2009, www.lacpa.org.lb/Includes/Images/Docs/TC/TC409.pdf.
36. Cole, A. An approach to the study of entrepreneurship, *J. Economic History*, 6(11), 183–84 (supplement), reprinted in Frederick C. Lane and Jelle C. Riesmersma, Eds., 1946.
37. Anand, R. "What type of an entrepreneur are you?" http://www.venturegiant.com/news-channel-282-what-type-of-an-entrepreneur-are-you.aspx.
38. Inspired by Maxwell, J.C. *The Ten Irrefutable Laws of Leadership.* Nashville TN: Thomas Nelson Publishers, 1998.

Chapter 5

New Venture Creation

5.1 Introduction

"Entrepreneurship is a marathon, not a sprint."

Statistics show that approximately 20,000 new businesses will start this week alone. Over 85% of those businesses will have sales of less than $1 million per year and have less than 20 employees five years after their establishment. Most of those businesses are mom-and-pop establishments, composed of restaurants, retail stores, coffee houses, etc. In this Guide, we will concentrate on those new ventures that depend on formal technologies and seek large amounts of external capital for growth. These new ventures exhibit the following characteristics.

New ventures derive their **idea mining** primarily from five sources:

1. Prior, or current, employment
2. Teaming with others
3. Market observations (spotting opportunities)
4. Deliberate searches
5. Accidental encounters

New ventures derive **opportunities** primarily from four sources:

1. New technology (disruptive, groundbreaking, etc.)—Published sources
2. New regulations; de-regulations—Trade associations

3. Improved customer services—Customer inquiries
4. Value-added practices—Reverse engineering

New ventures are derived from the following **ingredients**:

1. Idea and proof of principle
2. Technical knowledge by founder
3. Personal contacts
4. Physical/capital resources
5. Orders

5.2 Corporate Life Cycle

"The difference between failure and success is one more time."

Just like human beings, organizations progress through five predictable life cycle events, such as being born (inception), childhood (survival), fast growth period (expansion), maturity, and (perhaps) corporate death as postulated by Scott and Bruce.[1] The typical life cycle trajectory of a venture is shown in Figure 5.1.

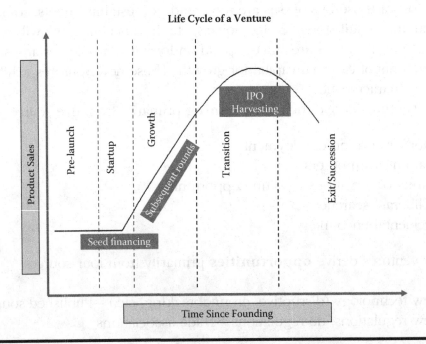

Figure 5.1 Life cycle of a venture—Similar to all living creatures, ventures display a predictable life cycle.

Small business ventures go through the life cycle because the business environment changes with time. The four most common business environment changes are:

1. Different competition
2. Market dynamics
3. Regulatory changes
4. Loss of a niche market or technology

5.3 Global Fortune 500 Annual Turnover

Obviously, only 500 companies comprise the annual Global Fortune 500 list, but it is illustrative of the rapidity of business changes throughout the world. In the 1950s, annual turnover (the number of places on the list that change as companies enter and exit the top 500 list) was moderate, and then lower through the 1960s and 1970s. Then, starting in the early 1980s, annual turnover rose to historically high levels, reaching new heights in the 1990s.

Every spring, *Fortune* magazine publishes its list of the largest public companies. This turnover provides a statistical barometer of business revenues. Table 5.1 lists the latest global 500 largest companies. These companies made the Global Fortune 500 in 2010, but fell off the list in 2011.[2]

5.4 Corporate Life Cycles

"Starting up is hard to do."

Startups face a daunting task to get off the ground. From the start, the founder must decide the strategy that will be pursued. There are two strategies available:

■ Entry strategies
  – First movers (first mover advantage)
  – Followers
■ Survival strategies
  – Specialists
  – Generalists

Table 5.1 Global Fortune 500 Companies that Exited Listing in 2011

Company Name	2010 Global arrival 500 rank 2011 exit	Country
Acer	487	Taiwan
Akzo Nobel	479	Netherlands
Alcatel-Lucent	461	France
Cathay Life Insurance	259	Taiwan
CEPSA	369	Spain
China Railway Construction	105	China
China State Construction Engineering	147	China
Chrysler Group	205	U.S.
Cie Nationale Portefeuille	480	Belgium
Cigna	459	U.S.
Co-operative Group	476	Britain
Danske Bank Group	454	Denmark
Doosan	489	South Korea
Evonik Industries	452	Germany
Groupama	420	France
GS Holdings	238	South Korea
Hanwha	321	South Korea
Hartford Financial Services	436	U.S.
Henkel	486	Germany
Hochtief	353	Germany
Kimberly-Clark	494	U.S.
Kirin Holdings	466	Japan
LG Display	440	South Korea
Marathon Oil	99	U.S.
Motorola Solutions	427	U.S.
Old Mutual	246	Britain
Petroplus Holdings	472	Switzerland
PPR	421	France
Premafin Finanziaria	493	Italy
Research In Motion	490	Canada
Samsung C&T	492	South Korea
Samsung Life Insurance	333	South Korea
Shanghai Automotive	151	China
Sinosteel	354	China
Standard Life	330	Britain
Tohoku Electric Power	488	Japan
U.S. Bancorp	477	U.S.
Wolseley	474	Britain

First movers (pioneers), as the name indicates, are the first entrants to the market. The classic example is a pharmaceutical company to introduce a new category of drugs immediately after obtaining FDA approval. All other pharmaceutical entrants will follow with a "variation of the theme" product. From a marketing standpoint, the first movers may dominate the largest market share for a prolonged period of time.

Followers may never gain much market traction, but are not as financially exposed as the first movers are. Current examples are car manufacturers Fisker and Tesla, who were pioneers in the introduction of electric cars, but were forced to declare bankruptcy in 2013. Perhaps follower companies, with the advantage of hindsight, may be more successful.

Specialists concentrate all their skills in one single niche market, develop a formidable core competency, and end up offering a family of superior products, at least for a while. If successful, specialists will be copied by followers, capitalizing on the specialists' initial success.

Generalists attempt to spread their skills across many market niches ("do not put all your eggs in one basket"). Generalists are most successful when the business environment is uncertain because their risk profile has been spread over a variety of market niches.

5.4.1 Greiner's Model of Organizational Growth

In 1972, Greiner published his seminal research on the evolution and revolution as organizations grow.[3] According to this model, organizations encounter a predictable series of problems (crises) that must be managed if the fledgling organization is allowed to survive, thrive, and mature in a complex and challenging environment.

The Greiner model consists of five stages of growth. To advance from one stage to the next, the organization must successfully manage and solve the corporate, financial, and technological problems seen in each stage, including the predictable "crises" typically associated with each stage. This is illustrated in Figure 5.2.

According to this model, startups are characterized by **stages**, defined as a number of predictable, discrete, and consistent growth phases. **Revolutions** are defined as those periods of substantial turmoil in organizational life. **Crises** are an important feature, defined as periods of relatively stable growth, interspersed with periods of chaotic and discontinued growth.

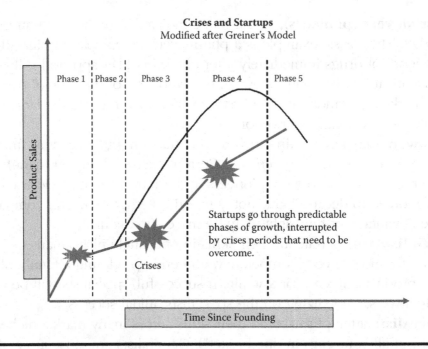

Figure 5.2 Crises and startups—Crises are inevitable periods of chaotic disruptions during organizational growth.

Startups are likely to experience the following crises during their growth life-cycle:

■ Crisis of ownership
■ Crisis of strategy, mission, and vision
■ Crisis of follow-on financing, sales, and marketing
■ Crisis of business model
■ Crisis of management turnover; role of founder

5.4.2 Churchill and Lewis Growth Model

In 1983, Churchill and Lewis published their influential framework on small business growth.[4] The authors categorized the problems of growth patterns in a systematic way. The startup capacity for growth is characterized by independence of action, different organizational structures, and varied management style, as shown in Figure 5.3.

The Churchill and Lewis model is helpful in anticipating the key requirements at the various growth phases, that is, the inordinate time and effort expended by owners/entrepreneurs during the startup period. This

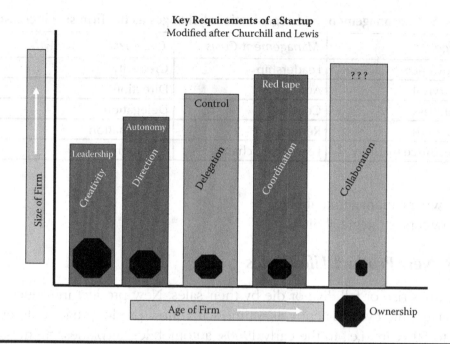

Figure 5.3 Key requirements of a startup—As the firm ages, ownership by founders declines, and delegation to experts becomes more critical.

is followed by the need for delegation and changes in managerial roles as the organizations grow, and the organizations become larger and more complex.

As Figure 5.3 indicates, when the size of the firm is plotted on the ordinate axis, against the age of the firm on the abscissa, a clearer view of the five stages of development, each characterized by size, diversity, and complexity as described by five management and organizations factors, as summarized in Table 5.2.

The authors further identified factors related to the enterprise and the owners as follows.

Four factors related to enterprise:

1. Financial resources
2. Personnel resources
3. Systems resources
4. Business resources

Four factors related to the owners:

1. Owners' personal goals
2. Owners' operational abilities

Table 5.2   Management and organizational changes as the firm size increases

Stage	Management Crisis	Organizational Growth
1 Existence	Leadership	Creativity
2 Survival	Autonomy	Direction
3 Success	Control	Delegation
4 Take-off	Red tape	Coordination
5 Resource maturity	Leadership change	Collaboration

3. Owners' managerial abilities
4. Owners' strategic abilities

5.4.3 New Product Life Cycles

Companies rise or fall, live or die by their sales. New product introductions also have a life cycle. There was a time when new products had a life cycle of 30 to 50 years (i.e., in the early 1900s: automobiles, airplanes, stainless steel, etc. as seen in Figure 5.4). By the mid 1950s, product life cycles were measured in less than a decade, see Table 5.3.

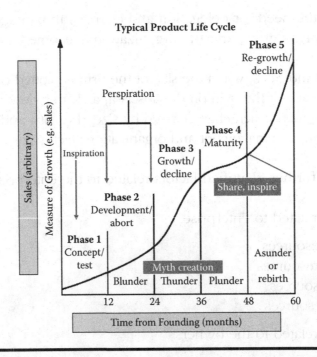

Figure 5.4   Typical product life cycle—Products, like ventures, display predictable life cycles.

Table 5.3 Estimated Decline in PC Sales

Device Type	2012	2013	2014	2017
PC (Desk-Based and Notebook)	341,263	315,229	302,315	271,612
Ultramobile	9,822	23,592	38,687	96,350
Tablet	116,113	197,202	265,731	467,951
Mobile Phone	1,746,176	1,875,774	1,949,722	2,128,871
Total	**2,213,373**	**2,411,796**	**2,556,455**	**2,964,783**

Source: Gartner, April 2013. Gartner Inc. is an information technology research and advisory company (www.gartner.com).

In the twenty-first century, many new products' life cycles are measured in months (i.e., smart phones, laptops, iPads, coronary stents, robotic surgery, medical imaging, etc.; Table 5.3).

As Table 5.3 indicates, worldwide devices (the combined shipments of PCs, tablets, and mobile phones) are on pace to total 2.4 billion units in 2013, a 9% increase from 2012, according to Gartner, Inc. Device shipments are forecast to continue to grow, reaching more than 2.9 billion units in 2017, but the mix of these devices will significantly change over the forecasted period.

As consumers shift their time away from their PC to tablets and smart phones, they will no longer see their PC as a device that they need to replace on a regular basis. Thus, the life cycle of the once-mighty PC is nearing its decline stage.

In general, new product life cycles follow a predictable pattern, as shown in Figure 5.5.

As an example, when PCs were first introduced to the public during the 1970s, their users kept those computers for an average of 14 years. As more computer power was added and prices decreased, computer manufacturers saw their competitors introduce newer PC generations. By the start of the twenty-first century, fewer and fewer PCs are kept more than 2 years, as seen in Figure 5.6.

Meanwhile, sales of PCs are plummeting as consumers find that disruptive technologies (tablets and smart phones) can perform many of the functions of modern desktop computers, but with more convenience and at competitive prices.

Moreover, with their high-definition touch screens, wireless Internet connections, and powerful processor chips, these mobile devices are more than

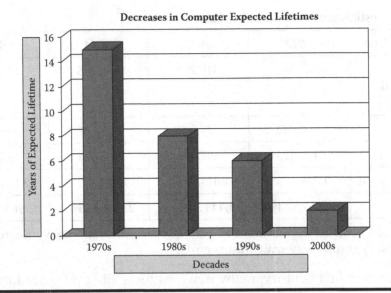

Figure 5.5   Decreases in computers—By the start of the second millennium, expected computer lifetimes are measured in a few months, not years.

adequate for reading e-mail messages, visiting websites, or even watching TV shows and movies.[5] Unlike PCs, consumers are routinely upgrading to newer, more powerful devices because of the two-year contracts offered by all major providers.

Plummeting PC sales herald big disruptions for traditional suppliers such as Hewlett-Packard, Dell, Microsoft, and Intel, whose microprocessors

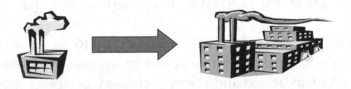

- You cannot spend "profit" or "net income." These are accounting figures only
- Cash is what you receive and can be used to pay bills or reinvested
- Cash flow does not equal net income. There are timing differences in accrual accounting when you record a transaction and when you <u>actually</u> receive the cash

Figure 5.6   Cash is king.

Table 5.4 Worldwide Devices Shipments by Segment (Thousands of Units)

Device Type	2012	2013	2014	2017
PC (desk-based and notebook)	341,263	315,229	302,315	271,612
Ultramobile	9822	23,592	38,687	96,350
Tablet	116,113	197,202	265,731	467,951
Mobile phone	1,746,176	1,875,774	1,949,722	2,128,871
Total	**2,213,373**	**2,411,796**	**2,556,455**	**2,964,783**

Source: Gartner, Inc. April 2013. www.Gartner.com

drive Windows and Mac computers. The mobile chip market is composed of giants such as Qualcomm Inc. and M Holdings. Even with these forces arrayed against them, PCs have not suddenly disappeared; with their comfortable keyboards and large storage capacity, PCs are nonetheless superior for tasks requiring large amounts of data input. Table 5.4 presents a side-by-side comparison of PC sales and tablets, as of April 2013.[6]

5.5 A Small Company Is Not a Little Big Company

This is the title of an influential paper in the *Harvard Business Review* by Welsh and White.[7] This 1981 business commentary is as relevant today as it was when originally published.

The authors argue that a traditional assumption among managers has been that small businesses can use essentially the same managerial principles "as the big boys," only on a suitably reduced scale. The basic assumption is that small businesses are like big businesses except that small companies have lower sales, smaller assets, and fewer employees.

Nothing could be further from the truth. Smallness creates what the authors call a special condition referred to as **resource poverty**. Resource poverty distinguishes small firms from their larger counterparts and thus requires critically different management styles and strategies. This can be summarized as shown in Table 5.5.

As we can see from Table 5.6, small businesses do not have the resources compared to big business, so how can they operate and survive? The

Table 5.5 "Resource Poverty" Applicable to Small Companies

Executive salaries	Represents a much larger percentage of overall costs
Human resources	Difficulty in attracting expensive but necessary talent Salaries vs. stock options—"skin in the game"
Business activities	Cannot afford large personnel expenditures in accounting, finance, marketing, sales, promotions, etc. Products launch expenses
External environment	Government regulations, industry standards Seasonal sales variations Insurance and banking needs Stakeholder demands
Internal environment	Cash flow management Reaching break-even point

owners of small businesses need to wear many hats and have multiple skill sets. From finance to accounting, from marketing to human resources, from operations to negotiations, the small-business person needs to understand all the elements of doing business.

So where and how does the small-business owner have the time to fulfill all these fundamental needs and functions, while at the same time create

Table 5.6 Disruptive Technology Examples

Old Technology	New Technology
Slide rules	Handheld calculators
Chemical photography	Digital photography
Movie theaters	VCR rentals
Vinyl records	CDs
Typewriters	Computer word processors
Open-chest coronary bypass surgery	Catheter-based coronary stents
Exploratory gastro-enteric surgery	Capsule endoscopy (pill that you swallow that transmits continuous images via telemetry)

the necessary innovations? Churchill and Lewis identified eight factors, four related to the enterprise and four related to the owner, as follows:

A. Company-related factors:
 1. Financial resources, including cash and borrowing power
 2. Personnel resources, relating to numbers, depth, and quality
 3. Systems resources, relating to information, planning, and control
 4. Business resources, relating to customer relations, market share, supplier relations, manufacturing, technology, and company position in its industry

B. Owner-related factors:
 1. Personal goals and business goals
 2. Operational abilities relating to marketing, inventing, producing, and distribution
 3. Managerial ability and willingness to delegate responsibility
 4. Strategic abilities in matching strengths and weaknesses of the company

5.5.1 Cash Is King

"Without cash there is nothing to manage."

A startup can survive as long as there is cash in the bank. The story is told and retold about the owner asking his accountant at the end of each month, "How much money do I have left in the bank?" Liquidity is a matter of life or death for the startup.

When you start a business, the vast majority of the time, your motivation is (and should be) first to break even and then to make a profit. Scott Allen, former About.com Guide, advises:[8]

One of the most important lessons entrepreneurs have to learn, often painfully, is that cash really is king. Not just paper money—but also *cash flow*. Simply put, it doesn't matter how much money will be coming in the future if you don't have enough money to get from here to there. Employees can't wait on paychecks until your customers pay. Your landlord doesn't care that you're talking to investors and will have the money in a couple of months. Suppliers may not be willing to extend your credit any further and you may not be able to purchase the goods you need in order to deliver to your customer and receive payment.

More businesses fail for lack of cash flow than for lack of profit. Why is this? Two main reasons:

1. Business owners are often *unrealistic* in predicting their cash flow. They tend to overestimate income and underestimate expenses.
2. Business owners often fail to *anticipate* a cash shortage and run out of money, forcing them to suspend or cease operations, even though they have active customers.

Let us start by differentiating between profitability and cash flow:

Profit is the difference between income and expenses. Income is calculated at the time the sale is booked, rather than when full payment is received. Likewise, expenses are calculated at the time the purchase is made, rather than when you pay the bill.

Cash flow is the difference between inflows (actual incoming cash) and outflows (actual outgoing cash). Income is not counted until payment is received and expenses are not calculated until payment is made. Cash flow also includes infusions of working capital from investors or debt financing. This is illustrated in Figure 5.7.

Cash flow is often calculated on a monthly basis, since most billing cycles are monthly. Most suppliers will typically allow somewhere close to thirty days to pay. However, in a cash-intensive business with a lot of inventory turnover, such as a restaurant or convenience store, it may be necessary to calculate on a weekly or even daily basis.

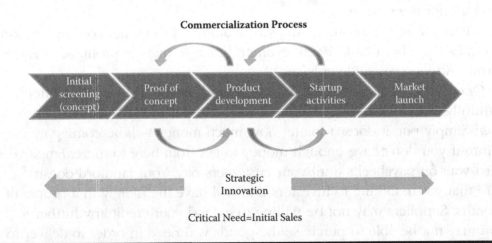

Commercialization Process

Initial screening (concept) → Proof of concept → Product development → Startup activities → Market launch

Strategic Innovation

Critical Need=Initial Sales

Figure 5.7 Innovation continuum—Innovation is your ticket to commercial and economic success.

5.6 Your Innovative Organization

"Innovate and change the world."

Most startups that rely on complex technology begin their business life by innovating a product or service. "Innovation is the ability to see change as opportunity, not a threat," as Steve Jobs famously quipped.

The process starts with an invention, and then progresses to innovation. **Invention** is the creation of new products or processes through new knowledge or from a combination of existing knowledge. **Innovation** is the initial commercialization of invention by producing or selling a new product, service, or process. **Commercialization** is the process of turning the invention, concept, or innovation into a product or service that can be sold in the marketplace. The entire process can be visualized as seen in Figure 5.8.

5.6.1 *Types of Innovation*

"Innovate or perish."

A culture of innovation can be the startup's primary source of competitive advantage. However, to create this culture you cannot merely hold a couple

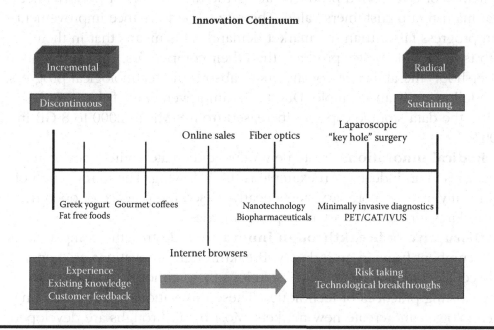

Innovation Continuum

Figure 5.8   Porter's value chain—The total value of an organization is greater than its component parts.

of meetings and pay lip service. You must execute. Some companies, such as Apple, are always innovating popular products, while most others are merely spectators in a contact sport.

Innovations are either (1) sustaining or (2) disruptive, as shown below:

1. Sustaining innovation; technologies
   - Continuous (incremental, evolutionary)
   - Discontinuous (radical, transformative)
2. Disruptive innovation; breakthrough innovations
   - Improve products in an unexpended fashion
   - Lower price
   - Enabling technologies

Incremental innovation is done by those actually implementing and using technologies and products as part of their normal activities. Most of the time, user innovators have some personal experience as motivation.[9] Incremental innovation is frequently observed within large established organizations, and is generally performed by intrapreneurs.

Sustaining innovations give customers the same value they already have, but with better attributes. Sustaining technologies improve the performance of established products, along the dimensions of performance that mainstream customers value. The rate of performance improvement can progress faster than the market demand. This means that in their efforts to provide better products than their competitors, companies often "overshoot" the ability of consumers to absorb the technological progress. The following is an example. Due to the improvements of the USB flash drive, the data storage capacity increased from 8 MB in 2000 to 8 GB in 2007.

Radical innovations create new value and create entirely new competencies in an industry. An example is the following. The introduction of minimally invasive coronary stents nearly erased the practice of open-heart procedures for coronary artery bypass surgeries.

Disruptive or **breakthrough innovations** destroy the competencies of incumbent firms in an industry. Breakthrough innovations are generally considered "out-of-the-blue" solutions that cannot be compared to any existing practices or techniques. These innovations employ enabling technologies and create new markets. Most breakthroughs are developed by R&D groups that often have not thought specifically about a particular commercial market application. These technologies originate on the supply side of the supply chain. Conventional wisdom says listen to the market, but breakthroughs come from labs that do not have what the customer wants in mind. These technologies are then pushed onto the consumer.[10] Examples include digital medical records, super-strong glass that keeps mobile device screens from breaking, and Starbucks leading the way for mainstream mobile payments.

The term **disruptive technologies** was coined by Clayton M. Christensen and introduced in his 1995 article "Disruptive Technologies: Catching the Wave,"[11] which he co-wrote with Joseph Bower. Currently, "disruptive" is a word used in the business and technology literature to describe innovations that improve a product or service in ways that the market does not expect, typically by lowering price or designing for a different set of consumers.

Disruptive innovations create a new (and unexpected) market by applying a different set of values. This is summarized in Table 5.6.

Innovative startups command a price premium in the financial markets. The price premium may be seen in the startup initial valuation, or may become apparent during subsequent financing rounds. Innovation is what gives startups their strength, staying power, and value. The concept of innovation continuum is shown in Figure 5.9.

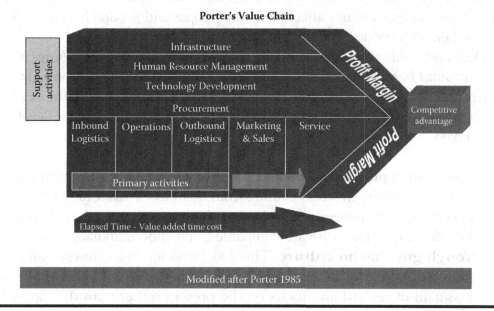

Figure 5.9 Value equation—How much "value" (benefits) are you delivering to your customers?

5.7 Creating Your Own Innovation Culture

"Culture eats strategy for breakfast."

An innovation culture is concerned with (1) the discovery of hidden opportunities and (2) the systematic commercial exploitation of proprietary technologies. As Gary Hamel, the Harvard business strategy guru stated, "Pursuing incremental improvements while rivals reinvent the industry is like fiddling while Rome burns."[12]

An **innovation culture** is one of the key drivers for the success—or failure—of a startup organization. A good, well-aligned culture can propel it to success. However, the wrong culture will stifle its ability to adapt to a fast-changing world. So, how do you attempt to understand your corporate culture? In addition, what steps can you take to create a strong corporate culture that will best support your organization's activities?

As founder, you must decide what type of organization culture your organization will follow. **Organizational culture** is the collective behavior of humans who are part of an organization and the meanings that the people attach to their actions. Culture includes the organization values, visions, norms, working language, systems, symbols, beliefs, and habits. It is also the pattern of such collective behaviors and assumptions that are taught to new organizational members as a way of perceiving, and even thinking and feeling. Organizational culture affects the way people and groups interact with each other, with clients, and with stakeholders.[13]

Deal and Kennedy[14] defined organizational culture as *the way things get done around here*. They created a model of culture that is based on four different types of organizations. They each focus on how quickly the organization receives feedback, the way members are rewarded, and the level of risks taken.

1. **Work-hard, play-hard culture**: This has rapid feedback/reward and low risk resulting in stress coming from quantity of work rather than uncertainty. High-speed action leads to high-speed recreation. Examples include restaurants, software companies, and shoe manufacturers.
2. **Tough-guy macho culture**: This has rapid feedback/reward and high risk, resulting in stress coming from high risk and potential loss/gain of reward and focus on the present rather than the longer-term future. Examples include police, surgeons, politicians, and sports figures.

3. **Process culture**: This has slow feedback/reward and low risk, resulting in low stress, plodding work, comfort, and security; stress that
   comes from internal politics and stupidity of the system; development
   of bureaucracies and other ways of maintaining the status quo; focus on
   security of the past and of the future. Examples include banks, insurance companies, teaching hospitals, and universities.
4. **You-bet-your-company culture**: This has slow feedback/reward and
   high risk, resulting in stress coming from high risk and delay before
   knowing if actions have paid off. The long view is taken, but then much
   work is put into making sure things happen as planned. Examples
   include aircraft manufacturers, oil companies, and startups.

5.8 Strategic Planning for StartUps

"Plans are nothing. Planning is everything." —Dwight D. Eisenhower

Strategic planning is the process used by an organization to visualize its
desired future and develop the necessary steps and operations to achieve
those aims. It directs managers to determine how they will be expected to
behave. In order to determine the direction of the organization, it is necessary to understand its current position and the possible avenues through
which it can pursue a particular course of action. Generally, strategic planning deals with at least one of three key questions:[15]

1. What do we do?
2. For whom do we do it?
3. How do we excel?

The key components of strategic planning include an understanding of
the firm's vision, mission, values, and strategies. Often a Vision Statement
and a Mission Statement may encapsulate the vision and mission.[16]

- **Vision:** Outlines what the organization wants to be, or how it wants
  the world in which it operates to be (an "idealized" view of the world).
  It is a long-term view and concentrates on the future. It can be emo-
  tive and is a source of inspiration. For example, a charity working with
  the poor might have a vision statement that reads, "A World without
  Poverty."
- **Mission:** Defines the fundamental purpose of an organization or
  an enterprise, succinctly describing why it exists and what it does to

achieve its vision. For example, the charity above might have a mission statement as "providing jobs for the homeless and unemployed."

■ **Values:** Beliefs that are shared among the stakeholders of an organization. Values drive an organization's culture and priorities and provide a framework in which decisions are made. For example, "Knowledge and skills are the keys to success" or "Give a man bread and feed him for a day, but teach him to farm and feed him for life." These example maxims may set the priorities of self-sufficiency over shelter.

■ **Strategy:** Strategy, narrowly defined, means "the art of the general"—a combination of the ends (goals) for which the firm is striving and the means (policies) by which it is seeking to get there. A strategy is sometimes called a roadmap, which is the path chosen to plow towards the end vision. The most important part of implementing the strategy is ensuring the company is going in the right direction, which is towards the end vision.

Unlike operational planning—which stresses how to get things done—and long-range planning—which primarily focuses on translating goals and objectives into current budgets and work programs—strategic planning is concerned with identifying barriers and issues to overcome. Managers are more likely to act on the assumption that current trends will continue into the future (steady-state management), while entrepreneurs need to anticipate new trends and possible surprises that represent both opportunities and threats.

5.9 Your Value Chain Analysis

The term "**value chain**" was first used by Michael Porter in *Competitive Advantage: Creating and Sustaining Superior Performance.*[17] The value chain analysis describes the activities that an organization must undertake and links them to its competitive strength and position.

The value chain concept revolves around the notion that an organization is more than just an agglomeration of machinery, equipment, facilities, technology, and human resources. Only when these support activities are aligned with primary activities will customers be persuaded to buy its products or services. The combination of all these factors becomes the source of competitive advantage. This is illustrated in Figure 5.10.

Notice the important distinction between primary and support activities. **Primary activities** are those directly involved with the creation or delivery

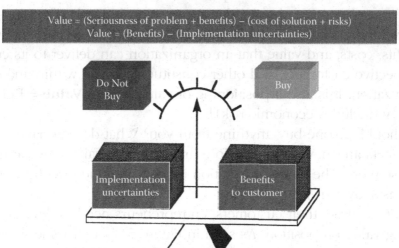

Figure 5.10   Value proposition—How compelling is your product/service to your customers?

of your product or service. **Support activities** help to improve effectiveness or efficiency of the operation. **Profit margin** is the ability of the organization to successfully deliver a product/service at a price that is higher than the combined costs of all the activities in the value chain. The numerical difference between price and cost is your profit margin.

You should perform your competitive advantage analysis within your value chain by:

- Analyzing which costs are related to every single activity
- Determining the optimal price of your product/service to your customer
- Identifying potential cost advantages you may have over your competitors
- Analyzing how your product/service potentially adds value (lower cost, higher performance, user-friendly, just in time delivery, etc.) to your customer's value chain

5.10 Your Value Proposition

"Killing two stones with one bird."

A **value proposition** is a promise of worth to be delivered and a belief from the customer that profit will be experienced. A value proposition can apply to an entire organization, or parts thereof, or customer accounts, or products or services. Creating a value proposition is a part of business strategy.[18] Developing a value proposition is based on a review and analysis of the benefits, costs, and value that an organization can deliver to its customers, prospective customers, and other constituent groups within and outside the organization. It is also a positioning of value, where **Value** = Benefits – Uncertainty (includes economic risk).[19]

Why should anyone buy anything from you? What do you have to offer? New products are "new" and therefore untested; generally, they cannot attract customers. The value proposition is best quantified by the value equation as shown in Figure 5.11.

In order to attract their customers, entrepreneurs need to develop a compelling value proposition. As seen in Figure 5.11, a customer would be persuaded to buy your new product/service if (a) the benefits outweigh the costs/risks, and (b) it solves a major serious problem. Figure 5.12 presents a step-wise process for establishing your value proposition.

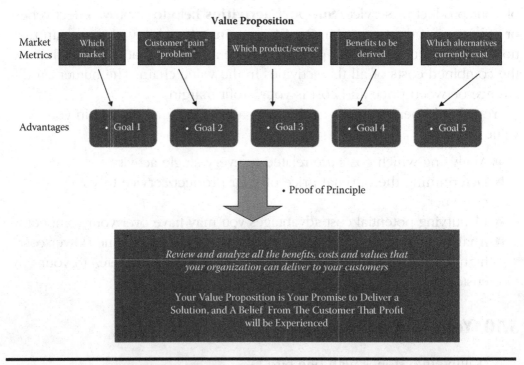

Figure 5.11 Generic SWOT analysis—A SWOT analysis is a powerful strategic planning tool to help you evaluate your competitive position vis-à-vis your intended objectives.

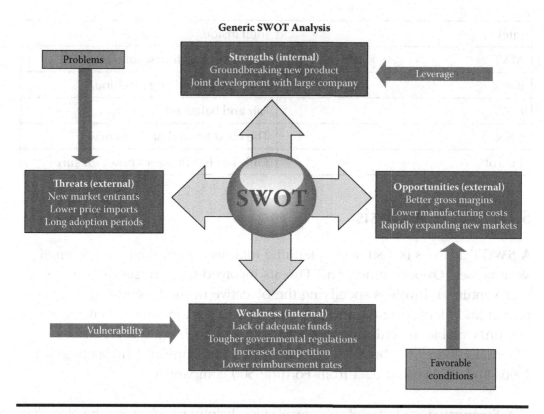

Figure 5.12   Early adopters—Early adopters will provide you with valuable "early warning" symptoms of desirability for your product offering.

Examples of value propositions (also called company slogans) include the following:

Coke	The pause that refreshes.
FedEx	The world on time.
E. F. Hutton	When E.F. Hutton talks, people listen.
American Express	Don't leave home without it.
Lexus	The passionate pursuit of perfection.
IBM	Global solutions for a small planet.
Apple	The power to be your best.
De Beers	A diamond is forever.
Visa	It's everywhere you want to be.

Intel	Intel inside.
AT&T	Reach out and touch someone.
BMW	The ultimate driving machine.
FOX	Fair and balanced.
CNN	The most trusted name in news.
Clairol	Only her hairdresser knows for sure.

5.11 SWOT Analysis

A SWOT analysis is a strategic planning tool used to evaluate the **S**trengths, **W**eaknesses, **O**pportunities, and **T**hreats involved in a project or in a business venture. It involves specifying the objective of the business venture or project and identifying the internal and external factors that are favorable and unfavorable to achieving that objective. The technique is credited to Albert Humphrey,[20] who led a research project at Stanford University in the 1960s and 1970s using data from Fortune 500 companies.

- **Strengths:** attributes of the organization helpful to achieving the objective.
- **Weaknesses:** attributes of the organization harmful to achieving the objective.
- **Opportunities:** *external* conditions helpful to achieving the objective.
- **Threats:** *external* conditions harmful to achieving the objective.

A generic SWOT analysis is presented in Figure 5.13.

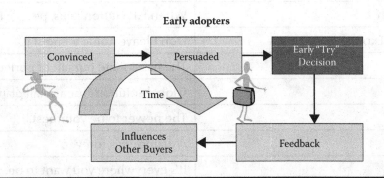

Figure 5.13 Technology adoption—Your job is to reach the growth period as soon as possible to ensure venture sustainability.

Your SWOT analysis should be conducted as follows:

- **STEP 1.** The present. List all your current strengths and weaknesses.
- **STEP 2.** The future. List all future opportunities and strengths.
- **STEP 3.** Your action plan. Address all four areas individually.
- **STEP 4.** Develop an operational plan, complete with specific tasks and dates of completion.

5.12 The Early Adopters

As an innovator, your first sales will likely be to an "early adopter." An **early adopter,** or trendsetter, is an early customer of a given product or technology. The term originated from the work of Rogers, *Diffusion of Innovations.*[21] Typically, this will be a customer who, in addition to using the innovative product or technology, will also provide considerable and candid feedback (beta site) to help refine its future product releases, as well as the associated means of distribution, service, and support.[22]

Early adopters are crucial to any new venture survival. Early adopters not only will provide some initial sales, but also will likely influence other buyers to "try out" the product or service. Early adopters are frequently called "cutting-edge" and "up-to-the-minute" customers. This is illustrated in Figure 5.14.

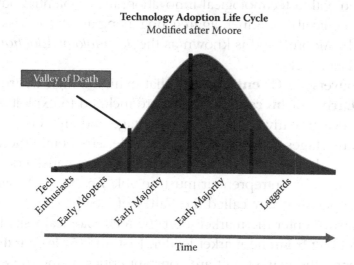

Figure 5.14   Valley of death—Also called the "funding gap." Do you have sufficient funds to survive the "Valley of Death"?

Figure 5.15 Valley of Death. Also called the "funding gap." Do you have sufficient funds to survive the "Valley of Death"?

When faced with a technological innovation, most potential buyers seem to react in a predictable manner, aggregating along an axis of risk aversion, as proposed by Moore[23] and is known as the *Technology Adoption Life Cycle*, as shown in Figure 5.15.

Potential buyers are (1) **enthusiasts** that exhibit a high desire to explore. (2) **Early adopters** (visionaries) are inclined to exploit an innovation to achieve an early competitive advantage and are very influential during the early stages of the adoption life cycle. Frequently there is a wide time gap in sales before the (3) **early majority** (pragmatists) decides to enter the market. The entrepreneur must be able financially to withstand this period, euphemistically called the Valley of Death. The (4) **late majority** (conservatives) enter the market after the technological risks have been mitigated, and buy from the market leader. Last, the (5) **laggards** (skeptics) may finally enter the market, but are constant critics. Using the 80/20 rule, laggards represent 80% of the complaints and criticisms experienced by the company.

5.13 Bridging the Valley of Death

"If you think you are going thru Hell, keep on going."
—Winston Churchill

The phrase "Valley of Death" was coined by Marczewski[24] to describe the missing link during the transition from existing technologies to the creation and acceptance of a compelling new market entry.

According to Frost and Sullivan, a consulting firm,[25] more than four out of five technologies developed globally never make it to the commercial world, due to their inability to cross the Valley of Death—the virtual chasm that separates the early market from technology demonstration in the mainstream market.

Often this is due to businesses and investors failing to understand the true market potential of a given technology platform and evaluate the risk-reward elements, as seen in Figure 5.16.

This is a period of exceptionally high financial vulnerability to the startup. The goal of the founder/entrepreneur is to financially and technically survive the Valley of Death and then win a niche foothold in the mainstream market as soon as possible. Examples include the first PC, the first fax machine, the Internet, the cell phone, the smart phone, robotic surgery, percutaneous heart valve repairs, etc.

References

1. Scott, M., & Bruce, R. Five stages of growth in small business. *Long Range Planning*, 20, 45–52, 1987.
2. From the July 23, 2012 *Fortune* magazine issue.
3. Greiner, L.E. Evolution and revolution as organizations grow. *Harvard Business Review*, 50(4), 1972.
4. Churchill, N.C., & Lewis, V.L., The five stages of small business growth. *Harvard Business Review*, May–June 1983.
5. Bray, H. For PC market, the numbers don't compute. *The Boston Globe*, Business, B7, April 15, 2013.
6. http://www.gartner.com/newsroom/id/2408515
7. Welsh, J.A., & White' J.F. A small business is **not** a little big business. *Harvard Business Review*, July–August 1981. Reprint No 81411.
8. http://entrepreneurs.about.com/od/beyondstartup/a/cashflow101.htm
9. http://en.wikipedia.org/wiki/Innovation
10. http://www.go4funding.com/Articles/Types-Of-Innovations.aspx

11. Christensen, C.M., & Bower, J.L. Disruptive technologies: catching the wave. *Harvard Business Review*, January–February 1995.
12. Hamel, G. Strategy as revolution. *Harvard Business Review*, July–August 1996.
13. http://en.wikipedia.org/wiki/Organizational_culture
14. Deal, T.E., & Kennedy, A.A. *Corporate Cultures*. New York: Perseus Book Publishing LLC, HarperCollins Publishing, 1982.
15. Renger, R., & Titcomb, A. A three step approach to teaching logic models. *American Journal of Evaluation*, 23(4), 493–503, 2002.
16. https://en.wikipedia.org/wiki/Strategic_planning
17. Porter, M.E. *Competitive Advantage: Creating and Sustaining Superior Performance*. The Free Press, 1985.
18. http://en.wikipedia.org/wiki/Value_proposition
19. Kaplan, R.S., & Norton, D.P. *Strategy Maps: Converting Intangible Assets Into Tangible Outcomes*. Cambridge, MA: Harvard Business Press, 2004.
20. Humphrey, A. SWOT analysis for management consulting. *SRI Alumni Newsletter* (SRI International), Dec. 2005.
21. Everett, R.M. *Diffusion of Innovations*. Free Press of Glencoe, Macmillan Company, 1962.
22. http://en.wikipedia.org/wiki/Early_adopter
23. Moore, G., Johnson, P., & Kippola, T. *How High-Tech Markets Develop. The Gorilla Game*. HarperCollins Publishers, 1999, pp. 21–46.
24. Marczewski, R.W. Bridging the virtual valley of death for technology. *R&D Scientist*, 11(2), 11–12, 1987.
25. Frost & Sullivan, Global Community of Growth, Innovation and Leadership Annual Congress GIL 2013, London, May 14, 2013.

Chapter 6

Financing Your Dream

CONGRATULATIONS!! If you have plowed through the previous chapters and are reading this chapter, you have decided to become an entrepreneur. Welcome aboard.

6.1 Naming Your Baby

You are now the parent of a new baby and, like all parents, you need to name your company and the products it will offer. As you will see, this is no easy task. Large and small companies alike have made embarrassing and costly mistakes.

Your company name will depend on two factors: (1) how much "gravitas" your company requires for customer acquisition. Gravitas, in this context, is the perception of your company's connectedness, funding, and relevance in its intended market(s), and (2) whether you are Business to Business ("B2B") or Business to Customer ("B2C"). Many B2C businesses depend upon advertising, potential to go "viral," perception, and social discovery for customer sales. Carefully consider how you intend to acquire customers, and then determine the importance of "name equity" in your customer acquisition strategies.[1]

6.1.1 Bad Company Names

An Egyptian airline, Misair, was very unpopular with the French because the name, when pronounced in French, meant "misery." The AMF Corporation had to change its name because AMF is the official acronym of the Australian Military Forces. Sew What, Hair Foyer, and Outerware Outhouse are other examples of poorly chosen company names.

6.1.2 Bad Product Names

The pharmaceutical giant Abbott Labs named its blockbuster anti-arthritis drug "Humira," which the company marketing department insists should be pronounced "hu-MARE-ah". General Motors had problems in Latin America when it introduced the Chevrolet "Nova." Literally translated in Latin, Nova means a new star, but in Spanish, it sounds like "no va," meaning, "does not go." Yves St. Laurent faced a storm of criticism in China when it named its new fragrance "Opium," an illegal drug.

6.2 Incorporating Your New Venture

Once you have selected your company's name, it is time (in conjunction with your lawyer) to select the firm's legal structure. Each structure has specific advantages and disadvantages, so your circumstances will dictate which legal structures are available, and which choice might be optimal. Your choices are shown in Figure 6.1.

Before we discuss the relative strengths and disadvantages of different corporate structures, Table 6.1 reviews the legal terminology associated with each entity type.

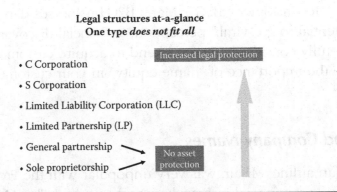

Legal structures at-a-glance
One type *does not fit all*

Increased legal protection

- C Corporation
- S Corporation
- Limited Liability Corporation (LLC)
- Limited Partnership (LP)
- General partnership
- Sole proprietorship

No asset protection

Figure 6.1   Legal structures—The six possible legal structures available to founders.

Table 6.1   Legal Terminology of Corporations

Term	C & S Corporations	LLCs	LPs & GPs
Owner	Shareholder	Member	General or limited partner
Executive management	Board chairman, CEO, CFO	Managers	General or limited partners
Incorporated in a state	Articles of incorporation	Articles of incorporation	Certificate of limited or general partnership
Governance documents	Bylaws	Operating agreement	LP & GP agreement

6.2.1 *Characteristics of Corporations*

Although corporate law varies by jurisdiction, the four core characteristics of the business corporation are: (1) legal personality, that is, perpetual existence, (2) limited liability, (3) transferable shares, and (4) centralized management under a board structure, as depicted in Figure 6.2.

6.3 Selecting the Optimal Legal Structure

To provide a frame of reference, Table 6.2 presents a simplified corporate entity comparison to simplify your discussions with attorneys, accountants, and other stakeholders. Selecting your legal structure is one of the most fundamental and critical early decisions.

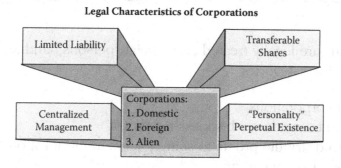

Figure 6.2   Legal characteristics—The four core characteristics of a business corporation.

Table 6.2 Simplified Corporate Entity Comparison

	C & S Corporations	LLC	LP & GP	Sole Proprietorship
Personal liability	None	None	None	Unlimited
Legal obligation	Officers and directors	Any member Any GP	Any GP	Sole proprietor
Managerial decisions	BOD Officers	Any member Any GP	Any GP	Sole proprietor
Ownership restrictions	Shareholder vote	Any member Any GP	Any GP	One sole proprietor
Startup documentation	Articles filed with state Annual meetings	Articles filed with state Annual meetings recommended	No state filing Annual meetings recommended	No state filing No annual meetings required
Interests transferability	Limited by securities laws	Unanimous or supermajority consent	Consent of all partners	Unlimited
Death of owner	Corporation continues	Dissolution in some states	Automatic dissolution	Automatic dissolution
Taxation of profits	Corporate tax rates; dividend taxes at individual rates	Individual tax rates	Individual tax rates	Individual tax rates of sole proprietor

Corporations are legally treated as artificial "persons," and are classified into the following categories:

■ For-profit corporations
  – Created to conduct a business for profit
  – Able to distribute profits to shareholders in the form of dividends
■ Not-for-profit corporations
  – Formed to operate charitable institutions, colleges, universities, religious groups, etc.

- Public corporations
  - Owned by shareholders
  - Securities listed and traded on stock exchanges
- Private corporations
  - Corporations owned by a few people and shares have no public market
  - Closely held corporations. Closely held corporations are entities funded and fully owned by founders as private companies. Since no "public" shareholders are involved, these organizations are also called "private."
- Corporations owned by one or only a few shareholders

6.4 Legal Powers of Corporate Executives

Publicly and privately held for-profit corporations confer **corporate titles** or **business titles** on company officials as a means of identifying their functions and responsibilities in the organization. The highest-level corporate executives are usually called "C-level" or part of the "C-suite," referring to 3-letter acronyms starting with "C" and ending with "O" (for "Chief ... Officer"); the traditional officers are Chief Executive Officer (CEO), Chief Operations Officer (COO), and Chief Financial Officer (CFO).[2]

Below are a list of important duties and responsibilities of corporate executives.

- Chief Executive Officer (CEO)
  - May also be Chairman of the Board
  - General fiscal responsibility and managerial supervision
  - Presides at all corporate meetings
  - Power to sign all documents
  - Reports annually to shareholders
  - The "buck stops here"
- President
  - May also be CEO
  - Day-to-day company supervision
  - Fixes compensation across entire entity
  - Implicit power to perform all acts authorized by the Board of Directors
  - Implements all strategic decisions
  - P&L (Profit & Loss) responsibilities

- ■ Chief Operations Officer (COO)
  - Responsible for the daily operation of the company
  - Routinely reports to the highest ranking executive, usually the CEO
  - Role is highly contingent and situational, as the role changes from company to company
  - In the manufacturing sector, the primary role of the COO is routinely one of operations management
  - Planning by prioritizing customer, employee, and organizational requirements
  - Maintains and monitors staffing levels
- ■ Chief Financial Officer (CFO)
  - Corporate officer primarily responsible for managing the financial risks of the corporation
  - Responsible for financial planning and record-keeping, as well as financial reporting to higher management
  - In some sectors the CFO is also responsible for analysis of data
  - Typically reports to the CEO and routinely attends Board of Directors meetings
  - Interfaces with auditors to provide quarterly and annual financial statements
  - Supervises accounting practices
- ■ Corporate Secretary
  - Responsible for the efficient administration of a company, particularly with regard to ensuring compliance with statutory and regulatory requirements
  - Ensures that decisions of the Board of Directors are implemented
  - Keeps corporate books
  - Communicates with shareholders to ensure that dividends are paid
  - Maintains company records, including directors and shareholder lists and annual accounts
- ■ Treasurer
  - Responsible for liquidity risk management, cash management, issues debt, and interest rate risk hedging, securitization
  - Oversight of pension investment management
  - Capital structure (including share issuance and repurchase)
  - Advises the corporation on matters relating to corporate finance
  - Oversees other financial activities such as insurance coverage

- Controller
  - Maintenance and audit of all financial records
  - Top managerial and financial accountant
  - Supervises the accounting department
  - Assists management in interpreting and utilizing managerial accounting information
  - Prepares operating budgets
  - Prepares tax filings with various local, state, and federal agencies
- Vice President
  - Assigned important and specific roles in corporate administration
  - Cannot legally bind the company without express or implied authority
  - May replace the President on the event of his or her death, resignation, or incapacity
  - Companies that use this title generally have large numbers of people with the title of Vice President with different functional responsibilities (VP Manufacturing, VP Marketing, VP Sales, etc.)

6.5 Creating Value for Your Stakeholders

"A job is a short-term solution to a long-term problem." —Robert T. Kiyosaki, author of *Rich Dad, Poor Dad*

In our society, there is a great reluctance to talk about how to make money, lots of money. Run out of money, and you will be forced to declare bankruptcy and see your dreams vanish. Universities are very good at teaching scholastic and professional skills, but are generally silent when it comes to money. Talking about money is not politically correct in academia.

As a fellow entrepreneur, I will discuss money with you without shame or guilt. My best advice to you: get rich and hire many good employees... after all, when was the last time you worked for a poor person? Table 6.3 discusses what money can do for everyone.

6.6 Employee Benefits

Your employee benefit program is an integral part of a total compensation package. Some benefits are government-mandated, while others are incentive-based to attract needed talent. Benefits required by government are

Table 6.3   Making Lots of Money Is Good (and Amazingly Altruistic)

Conventional Wisdom	Entrepreneurial Wisdom
The love of money is the root of all evil.	Money is the lifeblood of your organization.
Source of income = a good job	Source of income = value creation
Work in exchange for adequate money (salary).	Make money work for you. Multiplication is better than addition.
Write a good resume and get a job. Get along with everyone.	Write a business plan and raise lots of money.
Keep your nose clean.	Inspire greatness in your staff and create lots of multi-millionaires.
Pay taxes and take home the remainder.	Take full advantage of tax deductions and other legal investment incentives.
Work to provide for your family.	Work to provide gainful employment for many employees. You too, will gain in the process.

unemployment compensation, worker's compensation, Social Security, survivor benefits, disability benefits, and medical benefits. Figure 6.3 presents a summary of potential employee benefits.

Be aware that the Affordable Healthcare Act, once enacted, requires employers with 50 or more employees (or "full-time equivalents") to offer healthcare coverage or pay a $2,000 fine per employee (excluding the first 30 employees).

6.7 The 3 Fs, Angels, and VCs

At first, most entrepreneurs spend their own money, technical resources, and time in the yet-to-be formed company. At this point, the company is in the "dream" or "idea" stage and the entrepreneur is accumulating "sweat equity."

According to Investopedia, sweat equity "is the contribution to an innovative project or startup enterprise in the form of effort and toil." Sweat equity is the ownership interest, or increase in value, that is created as a direct result of hard work by the owner and is the preferred mode of building equity for cash-strapped entrepreneurs in their startup ventures because they

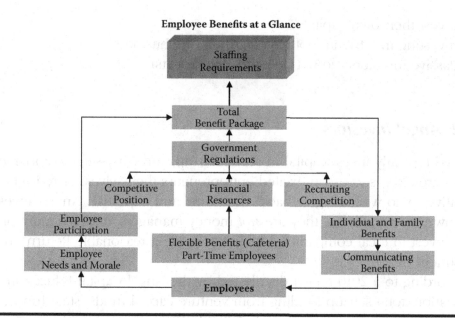

Figure 6.3   Employee benefits—Besides salary, what else can you offer your employees?

may be unable to contribute much financial capital to their enterprise. In the context of real estate, sweat equity refers to value-enhancing improvements made by homeowners themselves to their properties. The term is probably derived from the fact that such equity is considered to be generated from the "sweat of one's brow."[3]

For example, consider an entrepreneur who spends a year in his or her startup. After a year of developing the business and getting it off the ground, the entrepreneur sells a 25% stake to an angel investor for $500,000. The sale gives the business a valuation of $2 million (i.e., $500,000/0.25), of which the entrepreneur's share (sweat equity) is $1.5 million.

6.7.1 Friends and Family

Beyond sweat equity, most tech startups raise their first capital from friends and family. Friends and family financings are always the easiest to complete, often taking less than two months from start to finish. Friends and family rounds usually raise $25,000 to $150,000 in total.[4]

Friends and family investors share the following characteristics:

■ Not accredited
■ Unsophisticated in finance

- Invest their own capital
- Investing in a friend, not necessarily in the business
- Passive investors (do not demand Board seats)

6.7.2 Angel Investors

"Angels" typically invest capital in seed, startup, and early-stage companies. Angels are often successful excited entrepreneurs themselves, or retired executives who wish to "give back" their time and expertise. Angels invest their own money, that is, they are *not* money managers, and generally prefer to invest in local companies, looking to make a reasonable Return on Investment (ROI).

According to a 2010 report distributed by the Angel Capital Education Foundation, total startup funding from venture capital funds, state funds, and angel investors is approximately $20.8 billion annually. Surprisingly, friends and family contributed nearly three times that amount of capital to thousands of startups each year. With approximately $60 billion in startup funding from friends and family, entrepreneurs must consider this important option as they seek to launch new businesses.[5]

Angels are *accredited investors* (an SEC definition), which includes:

- Financial position:
  - Net worth: $1 million, or
  - Annual personal income: $200k or
  - Family income: $300K
- Assumptions:
  - Knowledgeable—capable of performing own due diligence
  - Can afford to lose the entire investment
- Implications:
  - Giving up regulated disclosure, but many are now part of Angel groups

6.7.3 Venture Capital

According to Wikipedia, venture capital (VC) is financial capital provided to early-stage, high-potential, high-risk, high-growth startup companies. A VC fund makes money by owning equity in the companies in which it invests, which usually have novel technologies or business models in high technology industries, for example, biotechnology, IT, or software. The typical VC investment occurs after the seed-funding round, frequently referred to as growth funding round (or Series A round). The VC seeks to generate returns through an eventual realization event, such as an IPO or a trade sale of the company.[6] VC is a subset of private equity.

One of the first steps toward a professionally managed VC industry was the passage of the Small Business Investment Act of 1958; this Act officially permitted the U.S. Small Business Administration (SBA) to license private "Small Business Investment Companies" (SBICs) to help finance and manage small entrepreneurial businesses in the U.S.

Before World War II, money orders (originally known as "development capital") were primarily the exclusive domain of wealthy individuals and families. Modern private equity investments began to emerge after World War II with the founding of the first two VC firms in 1946—American Research and Development Corporation (ARDC) and J.H. Whitney & Company.

ARDC was founded by Georges Doriot,[7] the "father of venture capitalism" (and former dean of Harvard Business School and founder of INSEAD),[8] with Ralph Flanders and Karl Compton (former president of MIT), to encourage private sector investments in businesses run by soldiers returning from World War II. ARDC was the first institutional private equity investment firm that raised capital from sources other than wealthy families, although it had several notable investment successes as well. ARDC is credited with the first trick when its 1957 investment of $70,000 in Digital Equipment Corporation (DEC) would be valued at over $355 million after the company's IPO in 1968 (representing a return of over 1200 times on its investment and an annualized rate of return of 101%).[9]

6.8 Principles of Raising Capital

The amount of money you plan to raise should be sufficient to accomplish key milestones that will either (1) make your startup self-sufficient or (2) enable you to raise additional capital at *a higher valuation*. Higher valuations enable management to keep a greater percentage of the company.

The entrepreneur needs to prepare for the due diligence process. Due diligence is the analysis and evaluation conducted by firms considering an investment in your company, and focuses primarily on (1) your management team, (2) the market opportunity, and (3) your technology, including intellectual property protection.

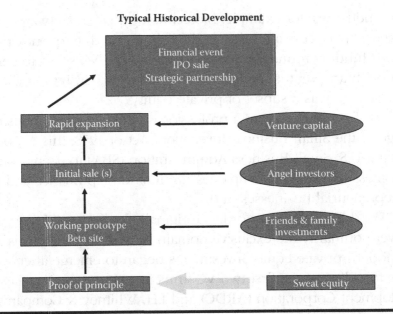

Figure 6.4 Typical historical development—Predictable steps undertaken by startups pre-funding.

Prepare a list of references and accomplishments of key management team members (including your scientific advisory board members) and your technology. Furthermore, have your patent firm prepare a status report on your patents, including a freedom to operate opinion, so you can verify that your products are proprietary and that you are not encumbered by the patents of others.

Most startups follow a predictable series of steps, prior to raising capital. Figure 6.4 summarizes the typical history of a startup.

The equity financing characteristics for startups is summarized in Table 6.4.

Table 6.5 summarizes the differences between debt and equity financing.

6.9 Persuasive Business Presentations

"Leadership is communication."

As a budding entrepreneur, you might as well get used to this: as an entrepreneur, you will be giving presentations until the cows come home. Moreover, persuasive presentations will be your trademark.

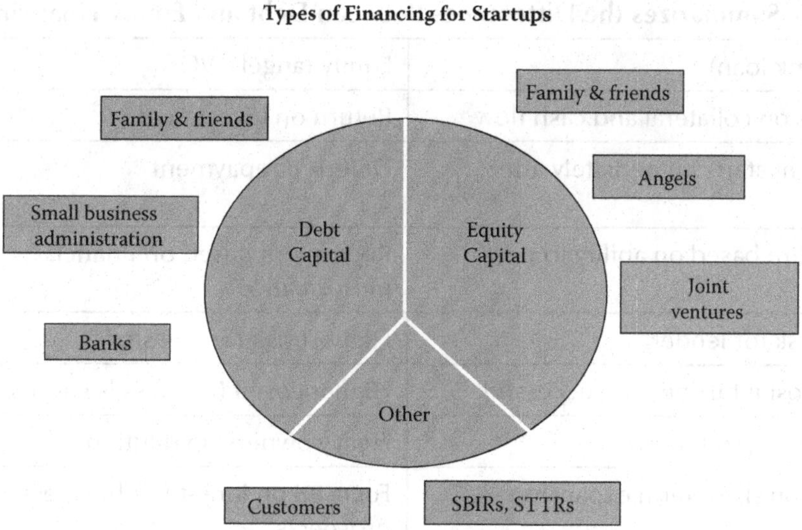

Types of Financing for Startups

Figure 6.5 Types of financing—You can finance your startup with (1) equity, (2) debt, or (3) other hybrid forms.

A **persuasive presentation** (speech) aims to get your audience to accept your business premise by prompting them to act, think, or feel in a desired manner, without coercion or force. Figure 6.6 presents the four cornerstones of persuasive presentations.

Pathos refers to presenting your reasons to believe in something, overcoming risks, natural apprehensions, perceived problems, etc. **Ethos** refers to your personal technical competence, goodwill, and dynamism to be trusted with investor's moneys. **Logos** are your set of rational, logical, and validated proofs. Last, **mythos** is the combined forces of ethical values,

Table 6.4 Equity Financing

Friends and family	Not necessarily "accredited" individuals
Typical round: $10–$100,000	"Passive" investment
	Personally interested in the technology
Angels	Accredited individual
Typical round: $50–$500,000	Expertise and personal investments
Increasingly as angel groups	Bets on the jockey, not the horse
VC	Professional investors
Typical round: $1–$5 million	LLP, GP
Looks for "exit"	Follow-on investments

Table 6.5 Summarizes the Differences Between Debt and Equity Financing

Debt (Bank loan)	Equity (angels, VC)
Emphasis on collateral and cash flow	Return on investment
Repayment starts immediately after funding	Deferred repayment
Debt return based on ability to pay	Repayment based on financial performance
Lowest risk for lender	Highest risk for investor
Lowest cost if business is successful	Higher cost if business is successful
No ownership dilution	Heavy ownership dilution
Focused on short-term expansion	Focused on long-term business prospects
Monitoring relationship	May demand Board plus upper management participation
Boilerplate documents	Complex documentation

industry beliefs, and national culture that may prompt investments in you and your company.

Persuasive speech is the most complicated form of verbal communication. It involves moving your audience to accept your premise from a position of deep skepticism/opposition to strongly/enthusiastically embracing your proposed solution based on its perceived benefits.[10] The entire sequence is shown in Figure 6.7.

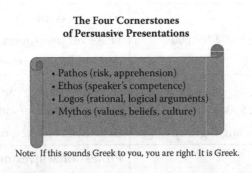

**The Four Cornerstones
of Persuasive Presentations**

- Pathos (risk, apprehension)
- Ethos (speaker's competence)
- Logos (rational, logical arguments)
- Mythos (values, beliefs, culture)

Note: If this sounds Greek to you, you are right. It is Greek.

Figure 6.6 The four cornerstones—Your basic forms of persuasion (prompting action without coercion).

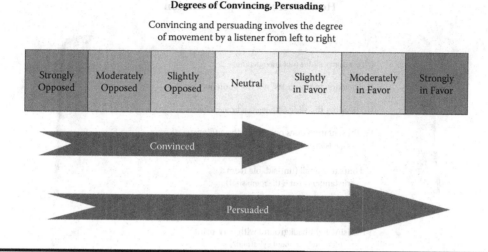

Figure 6.7   Degrees of convincing—Differences between convincing and persuading.

6.9.1 Rookie Mistakes

There is an old adage that goes, "Your presentation is 20% *what* you say, and 80% *how* you say it." Most rookie entrepreneurs tend to ignore their demeanor when making presentations, believing that their data "speaks for itself."

Another hurdle is the fact that most people become tongue-tied when placed in front of an audience. Most of us "freeze" when asked to give an important presentation. Did you hear the joke about the survey that asked aspiring entrepreneurs what are their three greatest fears in life? Their answers are as follows:

1. Fear of dying.
2. Fear of speaking in public.
3. Fear of dying while speaking in public.

Figure 6.8 presents a tongue-in-cheek list of don'ts for the young readers of this Guide.

6.10 Your Elevator Pitch

"I only had one superstition. I made sure to touch all the bases when I hit a homerun." —Babe Ruth

The "**Elevator Pitch**" derives its name from an apocryphal story: after having submitted a "teaser" document to a VC, and after having waited many

How to screw up a presentation

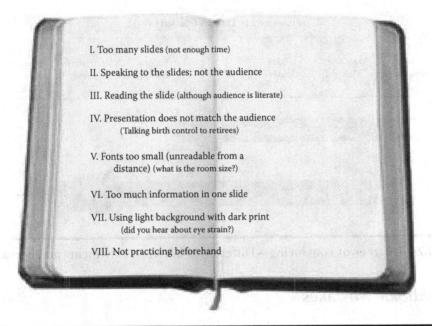

I. Too many slides (not enough time)

II. Speaking to the slides; not the audience

III. Reading the slide (although audience is literate)

IV. Presentation does not match the audience
    (Talking birth control to retirees)

V. Fonts too small (unreadable from a
    distance) (what is the room size?)

VI. Too much information in one slide

VII. Using light background with dark print
    (did you hear about eye strain?)

VIII. Not practicing beforehand

Figure 6.8 How to screw up—How to fail at your presentation in eight easy steps.

weeks to hear something from the VC, suddenly you get an unexpected phone call from the Managing Partner. "I am in the elevator going to a meeting. Tell me why I should fund your company now." The Partner has just asked you to answer three questions: Why me? Why you? Why now?

Guess what? You only have one chance of being funded. Thus, an Elevator Pitch must be a concise, carefully planned, and well-practiced description of your company that anyone should be able to understand in the time it would take to ride up three floors in an elevator. Like Babe Ruth, your pitch needs to touch these bases:

- A burning market need and your proposed solution to the burning need
- Your team and how they are uniquely qualified to manage the company
- How you will make money for your investors
- Memorable tagline/pitch closing

6.10.1 Must Haves

Your pitch should have a riveting **opening**; that is, grab the interest of your recipient. Your pitch should show **passion**—if you are not excited about your idea, no one else will be.

6.10.2 *Brief Descriptions*

Following the pitch be prepared to answer questions briefly. You must prepare a brief description of how the business is **different** from the **competition,** how you will **make money,** the **resources** you need from **investors,** and the **returns/payback** the investor can expect.

6.10.3 *Last Three Bits of Advice*

Always use the KISS principle—Keep It Simple, Stupid. Do not use techno-Latin, a language that only you understand. Highlight marketing advantages, not technical benefits.

6.11 Estimating StartUp Costs

Expenses incurred prior to the commencement of operations, startup expenses are incurred *after* the decision to proceed with the new business but *before* beginning operations, including:

- Business investigation expenses
- Organization costs
- Advertising
- Bank service charges
- Commissions
- Office and laboratory supplies
- Taxes (other than federal)
- Licenses, accounting fees, and legal fees
- Salaries and wages
- Utilities

6.12 Valuing You and Your Team

"It is better to invest in an 'A' team with a 'B' technology than in a 'B' team with an 'A' technology."

Just about everyone will tell you that management is the top factor for a successful startup. Furthermore, the entrepreneur represents 80% of the value given to teams. It is not surprising to hear seasoned investors speak of the

"ABCs" of their entrepreneurial investments. The "ABC" philosophy can be summarized as follows:

1. **Type "A".** The entrepreneur is technologically experienced, and has a successful record of running innovative companies. (Least risky)
2. **Type "B".** The entrepreneur is technologically experienced, but does not have managerial experience. (Average investment risk)
3. **Type "C".** The entrepreneur is neither technologically nor managerially experienced. (High risk)

Also, keep in mind that investors rarely provide you with all the required capital, preferring instead to invest in pre-determined stages (milestone-based). By staging their funding, investors retain the ability to abandon the project or re-value the company.

6.13 Valuing Your New Venture. (Calculating Pre-Revenue Valuation)

Valuation is the core determinant of return for investors. Unfortunately, seed and early-stage venture valuation creates the most contentious negotiations between the owner and investors. Aligning owner and investor expectations, particularly in the pre-revenue stages, is difficult and often leads to an impasse.

In this section, we will undertake a simple case of financing to illustrate how the process works and to demonstrate that valuation is more an art than a science and is ultimately determined by the marketplace and recalibrated annually.

6.13.1 Basic Calculations

Pre-Money valuation is the value of the company before any money is invested, while Post-Money valuation is the value after the money has been invested.

If an investment adds cash to a company, the company's post-money valuation will increase immediately after the investment.

External investors such as Angels and VCs will use Pre-Money valuation to calculate how much equity to negotiate in return for their cash infusion on a fully diluted basis.

Post-Money Valuation = Pre-Money Valuation + Money (investment)
Post-Money Valuation = Investment × (Total investment shares outstanding/shares issued for the new investment)
Pre-Money Valuation = Post-Money Valuation – new investment

6.13.2 Valuation Examples

Startup Capitalization Table				
Stockholders	*# Shares*	*$/Share*	*% Ownership*	*Enterprise $ Value*
Founders	1000	N/A	100	N/A

Financing Series "A" Objective = Negotiate raising $1 million for 200 newly issued Series "A" shares.

Series "A" per share value = $1 million/200 shares = $5/share
Implied Post-Money Valuation = $1 million * (1200/200) = $ 6 million
Implied Pre-Money Valuation = $6 million – $1 million = $ 5 million

Capitalization Table Following Series "A" Investment				
Stockholders	*# Shares*	*$/Share*	*% Ownership*	*Enterprise $ Value*
Founders	1000	5	1200/1000 = 83.3	5 million
Series "A" Investor	200	5	1200/200 = 16.7	1 million
Total	1200		100.0	6 million

Financing Series "B" Objective = Negotiate raising $2 million for 300 newly issued shares.

Series "B" per share value = $2 million/300 shares = $6.67/share
Implied Post-Money Valuation = $2 million × (1500/300) = $10 million
Implied Pre-Money Valuation = $10 million – $5 million = $5 million

Capitalization Table Following Series "B" Investment				
Stockholders	*# shares*	*$/share*	*% Ownership*	*Enterprise $ Value*
Founders	1000	6.67	1500/1000 = 66.7	6.67 million
Series "A" Investor	200	6.67	1500/200 = 13.3	1.33 million

Series "B" Investor	300	6.67	1500/300 = 20.0	2 million
Total	1500		100.0	10 million

A successful company can expect to have a series of investment rounds, ideally at Up-Round valuations (a higher Pre-Money valuation in each successive round). If the reverse were true, it would be called a Down-Round.

In our example, notice that the founders went from owning 100% of zero to owning 66.7% of a $10 million enterprise, that is, $6.67 million in equity. That is called value creation—get the picture?

References

1. Shapiro, J. http://thenextweb.com/entrepreneur/2012/04/22/before-naming-your-startup-read-this/.
2. http://en.wikipedia.org/wiki/Corporate_title
3. http://www.investopedia.com/terms/s/sweatequity.asp
4. http://www.angelblog.net/Startup_Funding_the_Friends_and_Family_Round.html
5. Hoeksema, A. http://blog.startupprofessionals.com/2010/08/friends-and-family-largest-startup.html.
6. http://en.wikipedia.org/wiki/Venture_capital
7. WGBH Public Broadcasting Service. Who made America?—Georges Doriot.
8. Ante, S.E. *Creative Capital: Georges Doriot and the Birth of Venture Capital.* Cambridge, MA: Harvard Business School Press, 2008.
9. Venture Impact: The Economic Importance of Venture Backed Companies to the U.S. Economy. NVCA.org. Retrieved 2013.
10. Lucas, S.E. Speaking to Persuade. Chapter 15. http://www.jdcc.edu/includes/download.php?action=2023&download_file_id=5274&action=2023&table_num=.

Chapter 7

Organizational Structure

7.1 Introduction

This chapter focuses on the crucially important task of building a new venture team. The point emphasized is that a firm's new venture team doesn't consist merely of its founders and its initial management and employees, but of its board of directors (if it is incorporated), its board of advisors, its lenders and investors (if applicable), plus all other professionals that compose the stakeholders.

The founders must also be aware of the "liability of newness." New ventures historically have a high propensity to fail. The high failure rate of the ventures is due primarily to three reasons:

1. The founders are incapable of adjusting to their new roles.
2. The firm lacks a "track record" with outside buyers and sellers.
3. Failure to assemble a talented and experienced founding team of risk-takers.

7.2 The Founding Team

"The first responsibility of the founder is to define reality."
—Mark DePree

In nearly 80% of startups, there is a founding team. Particularly in the early stages, the hours are excessive, the challenges are constant, and entrepreneurs find themselves in the company of a few other people who they will

---

An Effective Founding Team

- Managerial capabilities – skills, knowledge, experience
- Technical competencies
- Common vision
- Worked together previously
- At least one member with experience in the industry
- A network of industry contacts
- Expertise in the basic functional areas of the business
- Dedicated to the startup and able to endure any financial constraints.

Figure 7.1 An effective founding team—The eight ideal characteristics that define an effective founding team.

work insanely hard with the goal of "getting the ship to port." This initial founding team is the anchor of every startup, and is what will ultimately determine its destiny.[1]

What are the main characteristics of an effective founding team? Figure 7.1 summarizes the ideal characteristics of an effective founding team.

However, for you (the founder), establishing a founding team is easier said than done. Most seasoned executives are terrified of leaving their comfortable day jobs in a large organization. Perhaps "downsized" executives are your best bet for getting talent. Besides the founder, who actually composes the founding team? Figure 7.2 attempts to give objective answers.

Who is the founding team?

- The people who get to claim victory?
- The scapegoats for a failed business or investment?
- The reason investors make investments?
- The reason other people join the company?
- The reason the company succeeds or fails?

Figure 7.2 Who is the founding team?—If you are the founder, do you also have a founding team?

7.3 Hiring Your Executive Team

"As hire As. Bs hire Cs."

People, people, and more people. More than anything else, the hiring of your executive team will decide the ultimate fate of your company. In the author's experience, at the Board level (during our formative years), we spent more time dealing with executive team "issues" than any other subject.

In their executive team, entrepreneurs should select people whose talents are recognizably complementary to their own weaknesses. To select the optimal executive team, start by objectively understanding yourself, what you know, and what your weakest areas of expertise are. This is easier said than done because self-criticism is as rare as rain in the desert, particularly after you have climbed to the top of the ladder by talent, perseverance, and guts (Figure 7.3).

In hiring your executive team, you will need to decide between a heterogeneous and a homogenous group.[2]

The members of a heterogeneous executive team are diverse in terms of their abilities and experiences.

In contrast, the members of a homogeneous executive team are similar to one another in terms of their abilities and experiences.

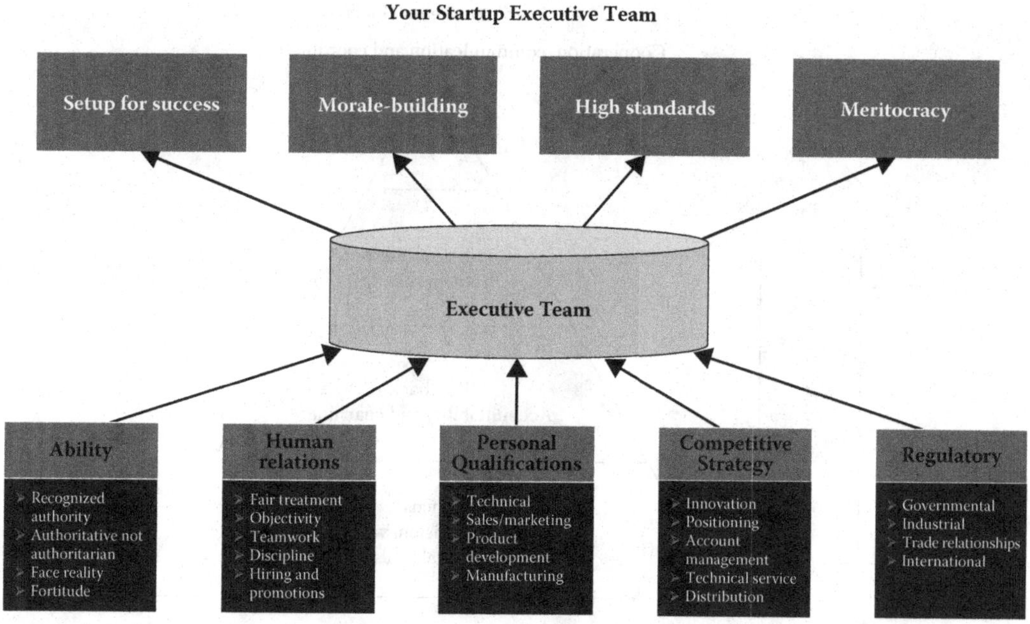

Figure 7.3 **Your startup executive team—Select executives with complementary skills to yours.**

Heterogeneous teams are traditionally favored by investors because (1) there is a shared intense effort that is required by a startup; (2) the loss of one member is less likely to result in startup abandonment; (3) the team concept allows expertise across major functional areas: marketing, finance, operations, sales, etc., and (4) a skilled team lends credibility to the startup and lowers risks to investors.[3]

7.3.1 Teamwork Not Titles

A real team is a small number of people with complementary skills, committed to a common vision. The vision includes (1) common purpose, (2) agreed approach, (3) performance standards, and (4) realistic goals for which they hold themselves mutually accountable.[4] In addition, goals at the executive team level are assigned based on individual skills (specialization), regardless of formal titles.

7.3.2 Team Discipline

The executive team must be goal oriented. Goals specify in advance what each member must accomplish, and periodically evaluate the degree to which members have met those goals. This is illustrated in Figure 7.4.

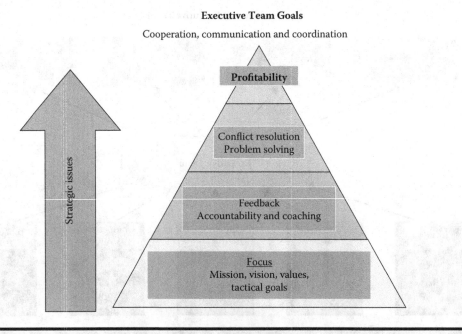

Figure 7.4 Executive team goals—Be crystal-clear about your goals to reach profitability.

7.3.3 *Focus on Outcomes, Not Activities*

"What gets measured gets done."

In most large companies, the vast majority of objectives are nothing more than activity-based goals. In his 1999 book, Douglas K. Smith laid out a guide for evaluating and realigning goals to achieve specific outcomes. "Activities are not objectives. Activities are how we achieve the objectives, the outcome-based goals."[5]

It is not enough to say, "This week I will visit a minimum of four customers." That is **activity**. Instead, "This week, out of the four customers, I will get a large purchase order from at least one." That is **outcome**. Figure 7.5 illustrates some important differences between activities and outcomes.

7.4 Too Many Chiefs, and Not Enough Productive Indians

"When a great executive meets up with a bad business, it is usually the business whose reputation remains intact." —Warren Buffett

When someone says there are *too many chiefs and not enough Indians* what they are really saying is that there are too many people wanting to be, or acting like, the boss and not enough people actually doing any work. The meaning of this expression is similar to the expression *too many cooks spoil the broth*.

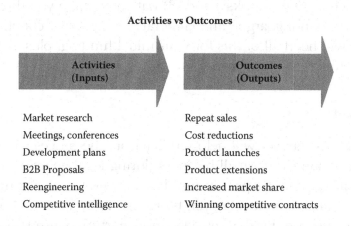

Figure 7.5   Activities vs. outcomes—Concentrate on outcomes (and be specific).

Do yourself one favor when starting your company. Initially, get the right people on board to get the product developed, delivered on time, and secure your initial critical mass of customers. Then, as you gain feedback from the market, figure out your next hiring needs. Don't worry about titles; instead, focus on building an insanely great product. Having too many chiefs and not enough Indians will burn a lot of dollars and will surely scare away potential investors in subsequent rounds.

7.5 Accountants and Lawyers

"Half of this game is 90% mental." —Yogi Berra

Accountants and lawyers are cost centers. Necessary, indispensable, and crucial, but cost centers nevertheless.

The founder must carefully select the proper accounting/legal team. By proper, I mean professionals who are familiar with the special circumstances surrounding startups. Engage only those individuals who are "specialized" in the intricacies of startups.

7.5.1 Accountants

Neither accountants nor lawyers academically "specialize" in startups. However, in practice, the representation of startups becomes a *de facto* specialization for both professions. For example, a small-business accountant will (1) help you with your business plan, (2) set up an inexpensive accounting system such as Quickbooks©, and (3) patiently guide you through the intricacies of accounting jargon and Generally Accepted Accounting Practices (GAAP). You will need all of this for your initial funding, plus all the necessary periodic reports to your investors.

7.5.2 Lawyers

The small-business lawyer will help you navigate the process of incorporation and representing your small business during legal negotiations. Perhaps the lawyer will agree to be paid in installments tied to specific milestones. Bob Loblow, a small-business lawyer, blogs that lawyers should be selected based on three criteria: local, right sized, and startup focused.[6]

The Walker Corporate Law Group listed the "top 10 reasons why entrepreneurs hate lawyers"[7] as follows:

10. They don't communicate clearly or concisely.
9. They don't keep me informed.
8. They are constantly over-lawyering.
7. They have poor listening skills.
6. Inexperienced lawyers are doing most of the work.
5. They spend too much time on insignificant issues.
4. They don't genuinely care about me or my matter.
3. Their fees are through the roof.
2. They are unresponsive.
1. They are deal-killers.

7.6 Managing Managers

"Be a coach, not a referee."

As founder, one of your most pressing goals should be to make sure that your managers have all the tools necessary to perform their jobs well. Managers need to understand not only the "what" but also the "why" behind your business plan. Give them a sincere sense of ownership in the organization and its future (Figure 7.6).

7.6.1 How Is Managing Managers Different From Managing Projects?

"We must all row in the same direction."

You will find that effective managers need to know less about "how" and more about "why." Managers should be evaluated primarily on team-based financial success, rather than products, schedules, or activities. Their opinions should carry extra weight along with responsibility and corresponding authority.

7.6.2 A Roadmap for Managing Managers Effectively

"Don't bring me problems; bring me solutions."

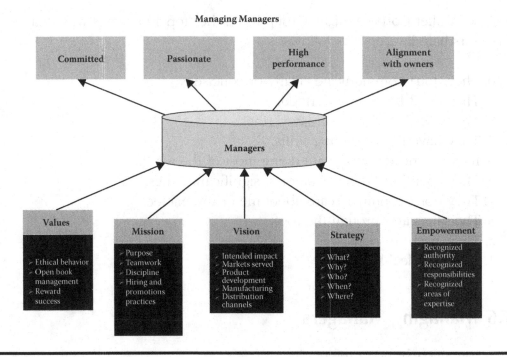

Figure 7.6   Managing managers—Empower your managers to perform their duties.

Bruce McGraw listed the following **important leadership skills for senior managers:**[8]

- Set the vision. It is essential that you communicate the long-term goals of your department or organization to your managers clearly and often. A shared vision provides the touchstone to help your managers make decisions and solve problems.
- Network with other managers and technical resources to get things done across your organization.
- Set straightforward, measurable objectives for each manager and project under your authority. Have short-term and long-term goals for each person who reports to you and reinforce those goals in monthly or quarterly meetings.
- Talented management-hiring gives effective feedback, and develops talent for the project work.
- Demonstrate accountability and hold team members and project managers accountable.
- Influence others up (your leaders), across (your peers), and down (your project managers and team).

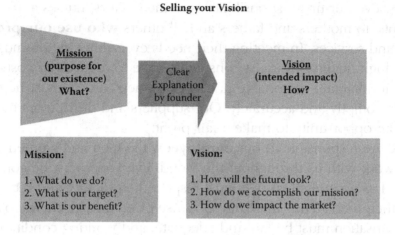

Figure 7.7   Selling your vision—Communicate your vision clearly and succinctly.

- Facilitate problem solving. It may be tempting to jump in and solve a problem yourself. You have been there and solved that problem successfully before. However, your managers need to learn and they need to put their own stamp on projects.
- Do not micromanage. Rather, offer advice and ask leading questions to help clarify a situation and your subordinate manager's options. You are also in a better position than before to break down barriers to solving problems by using your position and influence.
- Be a role model. Social learning theory, also called social cognitive theory, supports the idea that people learn new behaviors and change existing ones based not just on their experience, but also on their observations of significant others, such as senior managers. Your values, priorities, and even your mode of dress may be copied.

7.7 Selling Your Vision to Your Team

"Delegate to motivate."

Entrepreneurs must communicate their vision statement (Figure 7.7). For example, the Johnson & Johnson mission statement (the J&J "Credo")—chiseled into the wall of its New Jersey headquarters—was written in 1943 just before the company became publicly traded. Communicating it to J&J's multiple subsidiaries, so that each company lives it, is one of the J&J Chairman's major responsibilities.

We believe our first responsibility is to the doctors, nurses and patients, to mothers and fathers and all others **who use our products** and services. In meeting their needs everything we do must be of high quality. We must constantly strive to reduce our costs in order to maintain reasonable prices. Customers' orders must be serviced promptly and accurately. Our suppliers and distributors must have an opportunity to make a fair profit.

We are responsible to our **employees,** the men and women who work with us throughout the world. Everyone must be considered as an individual. We must respect their dignity and recognize their merit. They must have a sense of security in their jobs. Compensation must be fair and adequate, and working conditions clean, orderly and safe. We must be mindful of ways to help our employees fulfill their family obligations. Employees must feel free to make suggestions and complaints. There must be equal opportunity for employment, development and advancement for those qualified. We must provide competent management, and their actions must be just and ethical.

We are responsible to the **communities** in which we live and work and to the world community as well. We must be good citizens—support good works and charities and **pay our fair share of taxes.** We must encourage civic improvements and better health and education. We must maintain in good order the property we are privileged to use, **protecting the environment and natural resources.**

Our final responsibility is to our **stockholders.** Business must make a sound profit. We must experiment with new ideas. Research must be carried on, innovative programs developed and mistakes paid for. New equipment must be purchased, new facilities provided and new products launched. Reserves must be created to provide for adverse times. When we operate according to these principles, the stockholders should realize a fair return.

7.8 Fail as Fast as You Can

"Learning quickly why something failed is the key to rapid success."

When stakes are high (such as company survival), it makes sense to "fail fast." As an example, in the biotech field, startups need to go through a pre-described process of pharmaceutical product testing. The process, as described by FDA, consists of the following steps:

- Preclinical trials (involving laboratory animals)
- Phase I (involving a few healthy volunteers)
- Phase II (limited patients enrolled; compared to placebo)
- Phase III (larger patient population; compared to best available treatment)

Seek the truth. Costs increase exponentially at later stages. Be prepared to accept failure. If failure appears likely, it is best to cut your losses early. If your initial decision was based on the best available information available to you at the time, neither recriminate nor punish failure.

Convene your staff early to change direction. Failure is science's way of forcing you to change to a new direction. Conversely, if the drug seems safe and effective, be prepared to dive in headfirst. Reward your staff for a job well done (Figure 7.8).

7.9 Overcoming Organizational Inertia and Defenses

"Don't punish failure, reward success."

When the proper incentives do not exist for innovation, organizations develop inertia and defenses. Most employees are afraid to fail because

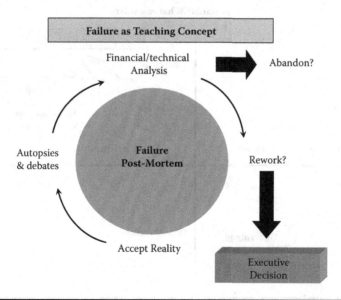

Figure 7.8 Failure as teaching concept—If it looks like you are going to fail, then fail as fast as you can.

failure likely can lead to dismissal. "Doing nothing" or "keeping your nose clean" is seen as a clever survival mode. It does not pay for individuals in large organizations to innovate.

In a breakthrough study of over 450 careers of successful men and women, Yeaple found that there are limited financial incentives to take personal risks when pushing an innovation.[9] This is shown in Figure 7.9.

Since innovation is the key to a startup, the founder must provide considerable upside to risk taking by authorized managers. As the old saying goes, "Show me an employee with a perfect record, and I will show you an employee who has not taken any chances."

7.9.1 Organizational Inertia

"Make stones out of stumbling blocks."

Overcoming organizational inertia is the extent to which the executive staff is motivated to change, learn, use, and accept a new direction through empowerment. **Empowerment** is the process of increasing the capacity of staff individuals or groups to make choices and to transform those choices into desired outcomes. It is accomplished through authority delegation.

Figure 7.10 summarizes the main factors at your control to overcome organizational inertia.

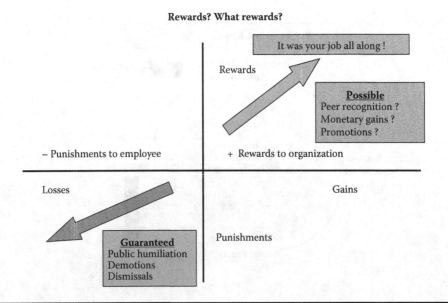

Figure 7.9 Rewards? What rewards?—Make sure you reward your innovators (and communicate how).

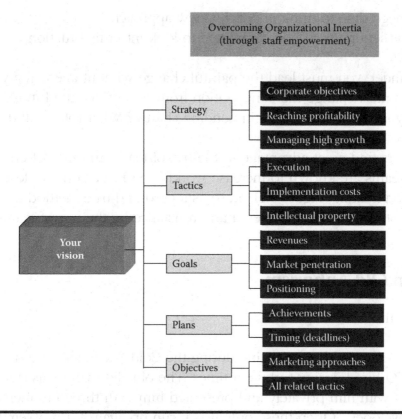

Figure 7.10 Overcoming organizational resistance—Empower your executives to overcome organizational inertia.

7.9.2 Organizational Defenses

"It is difficult to change things around here."

You can expect your organization to develop stubborn defenses against needed changes. Chris Argyris from Harvard Business School has been exploring this phenomenon for four decades.[10]

Organizational defenses are the pattern of obstacles presented to executives ("change agents") when a major managerial inflection point is reached during periods of fast growth or crisis. Some of the defenses take the following characteristics:

- We tried it before and it didn't work.
- There is nothing wrong with our current approach.
- Other companies in our field use the same techniques.
- We are too busy already.

- We need time to think about this new approach.
- Let us set up an ad hoc committee to look into this situation.

As founder, you must lead the painful change without alienating your team. You must change the conversation from "today" to "the future." How will everyone benefit from the impending change? What roles will they play in the new structure?

Founders find great advantage in the use of teams in times of crises because teams are known to increase workforce satisfaction and lead to better decision-making. Figure 7.11 presents a tried-and-true method for overcoming both organizational inertia and resistance by the use of teams.

7.10 An Old Aphorism

"The three envelopes."

An old joke (probably originating during the Cold War)[11] tells the story of a new CEO hired during turbulent times. The old CEO who was stepping down met with him privately and presented him with three numbered and sealed envelopes. "Open these only if you run up against a problem you don't think you can solve," he said.

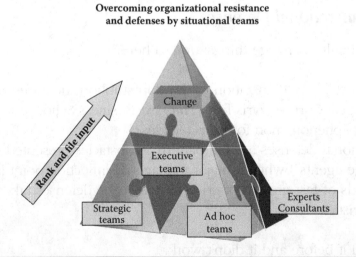

Overcoming organizational resistance
and defenses by situational teams

Figure 7.11 Overcoming organizational resistance and defenses—Establish situational teams in times of crises.

Things went along smoothly, but a few months later, sales took a downturn and the stock started to collapse. Then he remembered the three sealed envelopes. He went to his drawer and opened the first envelope. The message read, "Blame the old CEO."

The new CEO called a press conference and tactfully laid the blame at the feet of the previous CEO. Satisfied with his comments, Wall Street responded positively and the problem was soon behind him.

About a year later, the company was again experiencing a great dip in sales, combined with serious product problems. Having learned from his previous experience, the CEO quickly opened the second envelope. The message read, "Reorganize." This he did, Wall Street was appeased, and the company's stock quickly rebounded.

After several consecutive profitable quarters, the company once again fell on difficult times. The CEO went to his office, closed the door, and opened the third and last envelope.

The message said, "Write three envelopes."

References

1. Prive, T. Top 5 qualities of an unstoppable founding team. http://www.forbes.com/sites/tanyaprive/2013/05/31/top-5-qualities-of-an-unstoppable-founding-team/.
2. Building a New New Venture Team. Chapter 6. Prentice Hall and Pearson Digital Learning, http://prenticehall.com
3. Allen, K.R. *Launching New Ventures: An Entrepreneurial Approach,* 6th ed. Independence, KY: Cengage Learning, 2011.
4. Katzenbach, J.R. *Teams at the Top.* Cambridge, MA: Harvard Business School Press, McKinsey & Company, Inc., 1998.
5. Smith, D.K. *Make Success Measurable*! New York: John Wiley & Sons, 1999.
6. Loblow, B. http://www.bothsidesofthetable.com/2010/01/21/how-to-work-with-lawyers-at-a-startup/.
7. Walker, S.E. Top 10 reasons why entrepreneurs hate lawyers. http://venture-hacks.com/articles/hate-lawyers.
8. McGraw, B. Managing Managers Requires Good Leadership Skills, May 12, 2011. http://fearnoproject.com/2011/05/12/managing-managers-requires-good-leadership-skills/.
9. Yeaple, R.N. *The Success Principle.* New York: Macmillan/Spectrum, 1997.
10. Argyris, C. *Overcoming Organizational Defenses.* Allyn and Bacon, 1990.
11. https://www.facebook.com/Hahahalaughitout/posts/307976069255405

Chapter 8

Power Negotiations

"He who pays the piper calls the tunes."

The word **negotiation** is derived from the Latin "*negotiatus,*" which is past participle of "*negotiare,*" meaning "to carry on business." It is also related to "*negotium*" or "*necotium*" literally meaning "without leisure." A negotiation is a dialogue between parties, intended to reach an understanding, resolve a point of difference, or gain an advantage, to produce an agreement upon courses of action, or to satisfy various interests of parties involved in the process. Negotiation is a process where each party involved in negotiating tries to gain an advantage for themselves by the end of the process. Negotiation aims at a suitable compromise, in which parties can resolve their opposing interests.

Unlike litigation, business negotiation requires the consent of both parties. For this reason, negotiating involves the management of expectations. Both parties need to undertake negotiations with realistic expectations—not outcomes, but expectations. In mathematical terms, we can express this negotiation principle thusly:

$$\text{Negotiation Expectations} = \text{Satisfaction} = \Delta \,(\text{anticipation} - \text{reality})$$

8.1 Business Negotiations 101

"You can get much further with a kind word and a gun than you can with a kind word alone." —Al Capone

The term "negotiating in good faith" is more of a legal principle than a factual case. Negotiations occur primarily for three main reasons: (1) the need to agree on how to share or divide a limited resource, (2) to create something new that neither party could attain on their own, and (3) to resolve a problem or dispute between the parties.

Negotiation should be contrasted with **mediation**, where a neutral third party evaluates each side's arguments and attempts to reach a "suitable" or "fair" agreement between the parties. Negotiation should also be contrasted with **arbitration,** which resembles a legal proceeding. In arbitration, both sides make an argument as to the merits of their case and the arbitrator alone imposes the outcome.[1]

Closely related to negotiation is bargaining. **Bargaining** is a type of negotiation in which the buyer and seller of a good or service dispute the price or value that will be paid and the exact nature of the transaction, and eventually come to an agreement.[2]

8.1.1 Negotiation Wheel

When dealing with potential investors, as a founder you have more power than you think. Remember that like beauty, power is in the eyes of the beholder. Investors depend on you to come up with the innovations that will turn a profit for them. Without you, all they have is money. And they must invest that money or lose value as a result of yearly money devaluation.

Figure 8.1 depicts the negotiation wheel from a founder's perspective. The wheel assumes that this is your first "official" negotiation with an established investment consortium such as an Angel group. Notice that your individual objectives are at the center of the negotiating wheel. These objectives were the main reason why you undertook the risk of starting a fledgling company.

8.1.2 Negotiation Strategies

> "You got to know when to hold 'em and when to fold
> 'em." —Kenny Rogers

Negotiation theorists generally recognize two types of strategies: distributive and integrative (Figure 8.2).[3, 4, 5]

Three basic kinds of negotiators have been identified by researchers involved in The Harvard Negotiation Project.[6] These three types of negotiators are **soft bargainers**, **hard bargainers**, and **principled bargainers**.

Figure 8.1   Founder's negotiation wheel—State your objectives clearly and numerically.

- **Soft**. These people see negotiation as too close to competition, so they choose a gentle style of bargaining. The offers they make are not in their best interests, they yield to others' demands, avoid confrontation, and they maintain good relations with fellow negotiators. Their perception of others is one of friendship, and their goal is agreement. They do not separate the people from the problem, but are soft on both. They avoid contests of wills and insist on agreement, offering solutions and easily trusting others and changing their opinions.

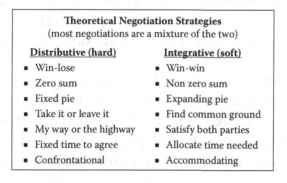

Figure 8.2   Theoretical negotiation—The two negotiation strategies at your disposal.

- **Hard**. These people use contentious strategies to influence, utilizing phrases such as "this is my final offer," "my way or the highway," and "take it or leave it." They make threats, are distrustful of others, insist on their position, and apply pressure to negotiate. They see others as adversaries and their ultimate goal is victory. Additionally, they search for one single answer, and insist you agree on it. They do not separate the people from the problem (as with soft bargainers), but they are hard on both the people involved and the problem.
- **Principled**. Individuals who bargain this way seek integrative solutions, and do so by sidestepping commitment to specific positions. They focus on the problem rather than the intentions, motives, and needs of the people involved. They separate the people from the problem, explore interests, avoid bottom lines, and reach results based on standards (which are independent of personal will). They base their choices on objective criteria rather than power, pressure, self-interest, or an arbitrary decisional procedure. These criteria may be drawn from moral standards, principles of fairness, professional standards, tradition, etc.

We will discuss distributive and integrative strategies at length in the subsequent paragraphs.

8.1.3 Distributive Negotiation[7]

Distributive negotiation is also sometimes called positional or hard-bargaining negotiation. It tends to approach negotiation on the model of haggling in a bazaar market. In a distributive negotiation, each side often adopts an extreme position, knowing that it will not be accepted, and then employs a combination of guile, bluffing, and brinksmanship in order to cede as little as possible before reaching a deal. Distributive bargainers conceive of negotiation as a process of distributing a fixed amount of value.

The term distributive implies that there is a finite amount of the thing being distributed or divided among the people involved. Sometimes this type of negotiation is referred to as the distribution of a "**fixed pie**." There is only so much to go around, but the proportion to be distributed is variable. Distributive negotiation is also sometimes called **win-lose** because of the assumption that one person's gain results in another person's loss. A distributive negotiation often does not involve people with a previous interactive relationship, nor are they likely to do so again in the near future.

8.1.4 Integrative Negotiation[8]

Integrative negotiation is also sometimes called interest-based or principled negotiation. It is a set of techniques that attempts to improve the quality and likelihood of negotiated agreement by providing an alternative to traditional distributive negotiation techniques. While distributive negotiation assumes there is a fixed amount of value (a "fixed pie") to be divided between the parties, integrative negotiation often attempts to create value in the course of the negotiation ("expand the size of the pie"). It focuses on the underlying interests of the parties rather than their arbitrary starting positions, approaches negotiation as a shared problem rather than a personalized battle, and insists upon adherence to objective, principled criteria as the basis for agreement.[9]

The word integrative implies some cooperation. Integrative negotiation often involves a higher degree of trust and the forming of a long-term relationship. It can also involve creative problem solving that aims to achieve mutual gains. It is also sometimes called **win-win** negotiation.

8.2 The Entrepreneur as Chief Negotiator

> "Wanting to be treated fairly because you are fair is like expecting the bull not to charge because you are a vegetarian."

Whether you want it, whether you fear it, the founder/entrepreneur *is* the startup chief negotiator. You will be up against much larger and stronger organizations, and so you must employ power negotiation. You operate a small startup and they are the largest Angel group in your state. However, contrary to popular fantasy, power is not dirty pool. Power is like weather: it can be either good or bad.

Power is the source of pressure you can exert to advocate your interests during a negotiation. Negotiating power provides you with the advantage/

leverage to gain a better outcome or preferred solution. For startups, power negotiation leverage is sought for two reasons:

- Equalization: to balance your opponent's initial advantage
- Enhancement: to secure needed outcomes

8.2.1 StartUp Negotiator Fears

"It is easier to go down a hill than up, but the best view is from the top of the hill."

Most startup founders/entrepreneurs are novices at the art of negotiation. For this reason, they dread the following three scenarios: (1) fear of looking foolish, (2) fear of the unknown, and (3) fear from past bad negotiating experiences. We all want to bat 1000. However, reality is that in business negotiation, if you are successful 50% of the time, you walk on water. Be prepared to fail (walk away) if it is a bad deal.

8.2.2 The Neophobia Syndrome

"If there is no wind, row."

Many startup founders/entrepreneurs exhibit neophobia. Neophobia is the initial aversion to something new or different. For inexperienced startup negotiators (neophytes), the author of this Guide recommends that you:

- Don't negotiate until you have created and established value—your differential competitive advantage.
- Don't discuss price until you are ready to finalize negotiations and close.
- Negotiate to close the deal as soon as possible.

Fortunately for the neophyte, there is a predictable negotiating cycle that can serve as a roadmap for your timetable, as shown in Figure 8.3.

The Negotiating Cycle	
Preparation	**Bargaining**
■ Set objectives	■ Get issues on the table
■ Fallback positions	■ Ask questions
■ Prioritize tradeables	■ Clarify
■ Set best, worst limits	■ Trade concessions
■ What if (????)	■ Memorialize
	■ Walkaway option

Figure 8.3 The negotiating cycle—Before bargaining, prepare, prepare, and prepare.

8.2.3 *If You Expect More, You Get More*

"Climate is what we expect, weather is what we get." —Mark Twain

One of the secrets behind power negotiating is that you initially ask your counterpart for much more than you expect to actually get. Henry Kissinger famously said, "Effectiveness at the conference table depends upon overstating one's demands."

Starting with high initial demands will later let you appear to be more cooperative because you are able to make larger concessions. In addition, asking for more than you expect automatically increases the perceived value of your offering, and lessens the possibility of a deadlock.

8.3 Knowing Your Negotiation "Counterpart"

"There is more to this than meets the eye."

In the real world, your negotiation "counterpart" or "partner" will fill a continuum from an experienced negotiator to an informal negotiator representing Friends and Family. Regardless of best intentions, on the opposite side of the negotiating table, your other "partner," "actor," or "adversary" is a human being and will act as such. In the context of negotiations, humans can be classified theoretically into five distinct camps, as shown in Figure 8.4.

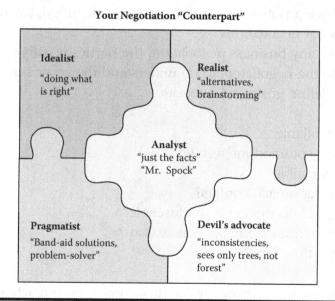

Figure 8.4   Your negotiating counterpart—Your counterpart will fill one or more of the listed categories.

Of course, your negotiating counterpart will not fit any single category all the time. As negotiations progress, each negotiator will predominantly fit into one of the five categories mentioned. However, be sure to insist that the other party appoint one negotiating leader that has the responsibility and the authority to close the deal.

If possible keep in mind the following rules:

- Don't assume anything. That only makes an ass out of you and me (ASS-U-ME).
- Learn all you can about each of the players on the other side.
- Keep in mind that negotiations fail in direct proportion to the number of attendees.
- In any negotiation, one attendee can easily kill the meeting by merely raising an objection.

8.4 Your First Negotiation

"Ride the horse in the direction it is going."

Since the first major negotiation of a founder/entrepreneur will likely be for a monetary investment in the startup, we will concentrate on that aspect of negotiations. Once a potential investor has been identified, the next step is engaging in formal negotiations.

At the start of any business negotiation, the parties usually start very far apart. It is up to the negotiators to (1) understand the other side's position, and (2) clearly define goals and interests.

- What is negotiable?
- Do we have common interests?
- Are we compatible?
- How do we approach conflict?
- What type of information is discloseable?
- How and when do we make concessions?
- Do we set a time limit?

As a rule, negotiations involve many tasks in order to reach a conclusion. It is nearly impossible to approach all the tasks, issues, and people at the same time. A good approach is to use "salami tactics." Like a large salami that is difficult to eat, the best way is to slice the salami into small eatable portions. To that end, some of the important discussion principles designed to facilitate a successful conclusion are listed next:

1. Separate process from substance.
2. Separate issues from people.
3. Break difficult issues into manageable tasks.
4. Assign a leader to solve each task.
5. Tasks are based on objective, verifiable references.
6. Write down the items discussed on a daily basis, and distribute to all concerned.

8.4.1 Aligning Interests

It is best to start by understanding and aligning each party's interests. The negotiators will then be able to assess (1) differences in interests, (2) differences in judgments about future directions, (3) differences in risk tolerance, and (4) differences in time preferences. Figure 8.5 summarizes some important starting interests.

8.5 Planning Your Negotiation Success

"The shortest distance between two points is not a straight line, but the path of minimum resistance."

In 1981, Fisher and Ury ignited a storm in the negotiations field with the introduction of their BATNA concept from the Harvard Negotiation Project.[10]

Aligning Interests

Entrepreneur wants	Investor/negotiator wants
■ Confidence on other parties' ability to close deal	■ Bridging valuation gap
	■ Establish deal exclusivity
■ Financial control	■ Qualify validity of business plan
■ Board composition	■ Timing of investment
■ Operational independence	■ Reps and warranties

Figure 8.5   Aligning interests—Start your negotiations by fully understanding each party's interests.

BATNA stands for "Best Alternative to a Negotiated Agreement." BATNA is the best outcome you could bring about if current negotiations fail and an agreement cannot be reached. If the negotiations are not delivering better value than your BATNA, then it is time to stop negotiating.

For example, if your startup company was valued at $1 million by Friends and Family in their seed round of financing, then your BATNA is $1 million (or higher) for any subsequent negotiations with Angels. This is your baseline.

Since their original introduction of BATNA, other acronyms have been introduced in an effort to facilitate comprehension and visualization of the negotiation process, such as:

- **WATNA** (Worst Alternative to a Negotiated Agreement)—WATNA is an option you may have to accept if the parties do not reach an agreement and your BATNA is unachievable. It is the worst outcome you would endure if negotiations fail. BATNA and WATNA are, in many ways, just two sides of the same coin. If you walk away, what is the worst that could happen? What are the consequences likely to be?
- **Target point**—The highest possible price (value) that could be attained for the business. Your preferred settlement talking the other party's limits and interests. Each item in the bargaining mix will have its own target, resistance, and starting point.
- **WAP** (Walk Away Price)—Price or value at which point the company will interrupt negotiations. It is understood that negotiations may be resumed in the future if a better offer is presented by the other party.
- **ZOPA** (Zone of Possible Agreement)—The ZOPA, in negotiations, describes the intellectual zone between two parties where an agreement can be met to which both parties can agree. Within this zone, an agreement is possible. Outside of the zone, no amount of negotiation will yield an agreement.[11, 12]

Figure 8.6 presents a summary of success planning terminology.

8.6 The Negotiation Agenda

"The agenda is an iron fist in a velvet glove." —G.D. Kieffer

Is it well accepted that "he who controls the agenda controls the meeting." Generally, the entrepreneur is at the mercy of the Angel's agenda when

---

Essentials of Success Planning

- Target Point (TP)
  – aspiration value
- Resistant Point (RP)
  – no further concessions are possible
- Zone of Possible Agreement (ZOPA)
  – Optimal for both parties
- BATNA Best Alternative to Negotiated Agreement
  – power to walk away
- WATNA Worst Alternative to a Negotiated Agreement

Figure 8.6   Essentials of success planning—Become familiarized with success planning terminology.

seeking funds. Angel groups have developed a unified agenda for meeting with all prospective clients, and thus start with a huge advantage, even before the negotiations take place. There are four distinct roles performed by an agenda:

1. Roadmap for meeting sequence
2. Names and titles of attendees
3. Communication of important issues to be discussed
4. Standard by which to measure success or failure of negotiations

8.6.1 *Preparing the First Draft of Any Document*

Every lawyer knows the strategic advantage of preparing the first draft of any legal document. The first draft is heavily tilted in favor of the drafter, placing the recipient in the disadvantageous position of having to make the document more balanced.

Whether it is a questionnaire, a Letter of Intent, or subsequent documents, the party preparing the first draft starts with an unquestioned advantage. If possible, the entrepreneur should insist on having his or her legal counsel write the crucial first draft of any document destined for a negotiation to ensure a more balanced approach.

The entrepreneur needs to be in constant touch with legal counsel in drafting the document, and not to assemble a committee. Remember that the easiest way to kill a deal is to get a committee working on it.

8.6.2 *Measuring Success*

"I will pay you anything you want, as long as I set the terms."

How do you know you have achieved success (or have failed)? What is your yardstick? In preparation for negotiations, assemble your team and set the following initial parameters, as shown in Figure 8.7.

8.6.3 *Negotiation Strategies*

Harvey Mackay, the nationally syndicated columnist and author of the *New York Times* #1 bestseller *Swim with the Sharks without Being Eaten Alive*,[13] has the following admonitions:

- Start with your negotiations with a problem-solving attitude.
- Never accept any proposal immediately, no matter how good it sounds.
- Never negotiate with yourself. You will furnish the other side with ammunition they might never have gotten themselves. Don't raise a bid or lower an offer without first getting a response.
- Never accept a deal with someone who has to "go back and get the boss' approval." Deal with decision makers only.
- Just because it may look nonnegotiable, doesn't mean it is. In business, "standard contracts" are not adhesion contracts.

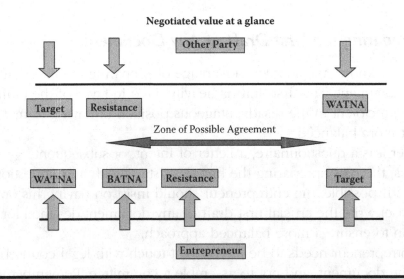

Figure 8.7   Negotiated value—Parameters used to set success expectations.

- A promise is a concession. It has a discount rate.
- Do your homework before you deal. Learn as much as you can about the other side. Instincts are no match for information.
- Beware the late dealer. Feigning indifference or casually disregarding timetables is often just a negotiator's way of trying to make you believe he or she doesn't care if you make the deal.
- A deal can always be made when both parties see their own benefit in making it.
- A dream is a bargain no matter what you pay for it. Set the scene. Tell the tale. Generate excitement. Help the other side visualize the benefits, and they will sell themselves.
- Make only deadlock-breaking concessions.
- Always strive for closure.
- Remember that satisfaction is the present value of future benefits compared to other choices.

The overarching negotiation strategy is the creation of future benefits, as depicted in Figure 8.8.

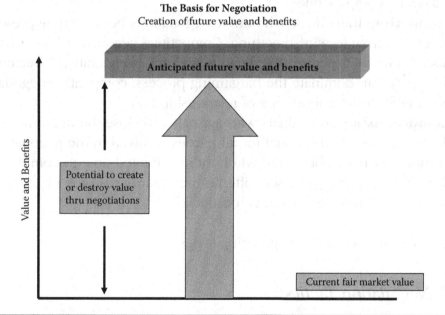

Figure 8.8 The basis for negotiation—Creation of future value as your basis for negotiations.

8.6.4 Negotiation Styles

Professor Shell[14] identified five styles/responses to negotiation. Individuals can often have strong dispositions toward numerous styles; the style used during a negotiation depends on the context and the interests of the other party, among other factors. In addition, styles can change over time, that is, negotiation styles are situational.

1. **Accommodating**: Individuals who enjoy solving the other party's problems and preserving personal relationships. Accommodators are sensitive to the emotional states, body language, and verbal signals of the other parties. However, they can feel taken advantage of in situations when the other party places little emphasis on the relationship.
2. **Avoiding**: Individuals who do not like to negotiate and don't do it unless warranted. When negotiating, avoiders tend to defer and dodge the confrontational aspects of negotiating; however, they may be perceived as tactful and diplomatic.
3. **Collaborating**: Individuals who enjoy negotiations that involve solving tough problems in creative ways. Collaborators are good at using negotiations to understand the concerns and interests of the other parties. However, they can create problems by transforming simple situations into more complex ones.
4. **Competing**: Individuals who enjoy negotiations because they present an opportunity to win something. Competitive negotiators have strong instincts for all aspects of negotiating and are often strategic. Because their style can dominate the bargaining process, competitive negotiators often neglect the importance of relationships.
5. **Compromising**: Individuals who are eager to close the deal by doing what is fair and equal for all parties involved in the negotiation. Compromisers can be useful when there is limited time to complete the deal; however, compromisers often unnecessarily rush the negotiation process and make concessions too quickly.

Figure 8.9 shows the five negotiating styles.

8.6.5 Negotiation Tactics[15]

Tactics are always an important part of the negotiating process. However, tactics do not often jump up and down shouting, "Here I am, look at me."

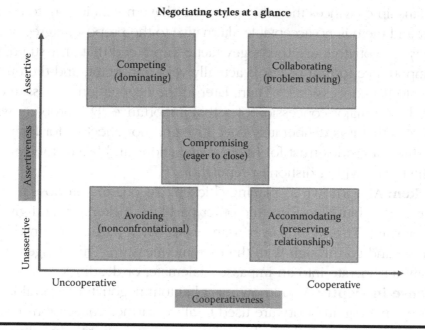

Figure 8.9 Negotiating styles—Your negotiating counterpart will likely fit one of these styles.

If they did, the other side would see right through them and they would not be effective. Frequently they are subtle, difficult to identify, and used for multiple purposes. Tactics are more frequently used in distributive negotiations and when the focus is on taking as much value off the table as possible. Many negotiation tactics exist. Following are a few commonly used tactics.

Auction: This bidding process is designed to create frenzied competition.[16] When multiple parties want the same thing, pit them against one another. When people know that they may lose out on something, they will want it even more. Not only do they want the thing that is being bid on, they also want to win, just to win. Taking advantage of someone's competitive nature can drive up the price. An auction does not work when your counterpart demands a "no shop" provision.

Brinksmanship: Secretary of State John Foster Dulles under the Eisenhower administration coined the term brinksmanship during the Cold War. One party aggressively pursues a set of terms to the point at which the other negotiating party must either agree or walk away. Brinksmanship is a type of "hardball" approach to bargaining in which one party pushes the other to the "brink" or edge of what that party is willing to accommodate. Successful brinksmanship convinces the other party they have no choice but to accept the offer and there is no acceptable alternative to the proposed agreement.[17]

Bogey: Negotiators use the bogey tactic to pretend that an issue of little or no importance to him or her is actually very important, and then use that ruse to get concessions.[18] Then, later in the negotiation, the issue can be traded for a major concession of actual importance. The problem with the bogey is that it is deliberately deceptive and not effective for long-term relationships. It erodes trust for future negotiations and leaves the party using the bogey with a dishonest reputation.

Chicken: Also known as playing chicken. Negotiators propose extreme measures, often bluffs, to force the other party to chicken out and give them what they want. This tactic can be dangerous when parties are unwilling to back down and go through with the extreme measure. This dangerous tactic can easily degenerate into an impasse, stalemate, or deadlock.

Defense in Depth: Also known as phantom negotiator. Several layers of decision-making authority are used to allow further concessions each time the agreement goes through a higher level of authority. In other words, each time the offer goes to a decision maker, that decision maker asks to add another concession in order to close the deal. For this reason, it is recommended that you negotiate with your counterpart's decision maker.

Deadlines: Give the other party a deadline forcing them to make a decision. This method uses time to apply pressure to the other party. Deadlines given can be actual or artificial. Deadlines work well when buying a car or in real estate transactions, but are very treacherous in a business negotiation setting.

Flinch: Flinching is showing a strong negative physical reaction to a proposal. Common examples of flinching are gasping for air or a visible expression of surprise or shock. The flinch can be done consciously or unconsciously. The flinch signals to the opposite party that you think the offer or proposal is absurd in hopes the other party will lower their aspirations. Seeing a physical reaction (body language) is more believable than hearing someone saying, "I'm shocked."

Good Guy/Bad Guy: Also known as good cop/bad cop. The good guy/bad guy approach is typically used in team negotiations where one member of the team makes extreme or unreasonable demands, and the other offers a more rational approach. This tactic is named after a police interrogation technique often portrayed in the media. The "good guy" will appear more reasonable and understanding and, therefore, easier to work with. In essence, it is using the law of relativity to attract cooperation. The good guy will appear more agreeable relative to the bad guy. This tactic is easy to spot because of its frequent use in Hollywood movies.

Highball/Lowball: Depending on whether selling or buying, sellers or buyers use a ridiculously high or ridiculously low opening offer that will never be achieved. The theory is that the extreme offer will cause the other party to reevaluate his or her own opening offer and move close to the resistance point (as far as you are willing to go to reach an agreement). Another advantage is that the person giving the extreme demand appears more flexible as he or she makes concessions toward a more reasonable outcome. A danger of this tactic is that the opposite party may think negotiating is a waste of time.

The Nibble: Nibbling is asking for proportionally small concessions that have not been discussed previously just before closing the deal. This method takes advantage of the other party's desire to close by adding "just one more thing."

Snow Job: The "snow job" consists of giving an opponent so much information (a blizzard) that he or she gets bogged down in trivia.[19] Negotiators overwhelm the other party with so much information that he or she has difficulty determining which facts are important and which facts are diversions. Negotiators may also use technical language or jargon to mask a simple answer to a question asked by a non-expert.

8.7 Laws of Power Negotiation

"Power is 20% given and 80% taken."

From the startup founder/entrepreneur perspective, their negotiation counterpart is always larger, more aloof, and capable of putting them in bankruptcy. In addition, the negotiating partners, be they customers, angels, venture capitalists, private equity, etc., know that fact very well and usually exploit it to their advantage.

A **power negotiator** can be defined as an actor with the situational ability to dominate and control the outcome of a negotiation. The high power negotiator can unilaterally satisfy the needs, desires, and aspirations of the other party. Thus, the founder/entrepreneur must be capable of claiming and creating future value to their counterpart.

The guru of power negotiating is Roger Dawson.[20] He emphasizes that in extended negotiations you will frequently encounter impasses,

Table 8.1 The three "situations" likely to be encountered during negotiations

Impasse	Stalemate	Deadlock
In complete disagreement on one issue, and it threatens to derail the negotiations.	Negotiations continue, but parties incapable of making progress.	Issues so intractable that there is no point in continuing negotiations.

stalemates, and deadlocks. What is the difference among these? Table 8.1 clarifies the differences.

In an **impasse**, you may recommend temporarily bypassing the offending issue, and reach agreement on other less contentious issues. An impasse is not always caused by world-ending issues or great matters of microeconomic uncertainty, but in many cases are the result of personality clashes or an inability to make a decision. An impasse usually requires an icebreaker to re-start negotiations.

You may decide to make the first move in breaking an impasse by changing emphasis from a competitive to cooperative, problem-solving mode by involving engineers with engineers, operations managers with operations managers, lawyers with lawyers, etc.

In a **stalemate**, you may want to change the dynamics of the negotiation by altering one or more of the issues, and volunteering methods of mutual risk sharing. Negotiation stalemates can be broken by two broad strategies:[21] (1) clarifying needs, interests, and issues, and (2) using a credible and achievable deadline.

Deadlocks are infrequent, but may require a third-party mediator. Roger Dawson also recommends attempting to resolve minor issues first, and then moving up the ladder toward the most controversial issues. Deadlock may not mean "dead"; it may mean one more stage of the bargaining process. If the parties are convinced they both have something to gain, they will continue to negotiate.

8.7.1 The Chicken or Egg Cycle

"Time is the shortest distance between two points."

Investors, buyers, and strategic partners know that entrepreneurs are under crushing time-related pressures to raise money, sell their products, or reach

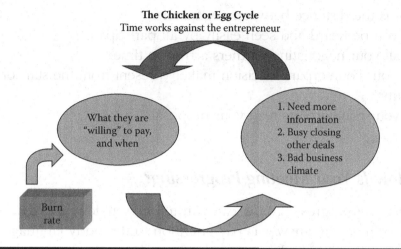

Figure 8.10 The chicken or the egg—Is time being used against you as a negotiating weapon?

a suitable agreement. Time is the entrepreneur's most implacable enemy. Remember that the party under the greatest time pressure will lose.

The other party frequently uses time as a pressure tactic, as depicted in Figure 8.10.

8.7.2 The Negotiating Table

"You are where you sit."

One of the most overlooked aspects of negotiation is the negotiating physical environment. Negotiations occur at the investor facilities. As in baseball, investors have the home turf advantage. They have the advantage of deciding on the location of the meeting room, the relative elevation of the participants, visual aids setting, access to files and archives, and importantly conference table configuration and seating arrangements.

The table configuration and seating arrangements convey important clues to the attitude of your negotiation counterparts. Win/lose attitudes of your negotiation counterpart are transmitted by their selected table configuration. Does their table arrangement promote trust or power? Before starting the meeting, ask yourself the following questions:

■ Were you asked to seat at one end of the table conveying an aura of authority?
■ Is the table large enough to seat all participants?

- What is the distance between negotiators?
- Can you be heard and seen easily by all participants?
- Did all your negotiating partners arrive on time?
- Was your counterpart "decision maker" present from the start of the meeting?
- Did your counterparts read your proposal?

8.7.3 How Is Your Meeting Progressing?

As the meeting progresses, how can you tell how much progress (if any) you are making? The answer is pay attention to the body language of your counterparts. Do they look interested or bored? Do they ask pertinent questions or are they skeptical? Are they looking at you in the eye or are they looking down or away from you? Are they sitting upright or slumping?

Figure 8.11 depicts the body language between the entrepreneurs (sitting on the right side of the table), and their negotiating counterparts. In this imaginary meeting, things are not going well for the entrepreneurs. If you are unlucky enough to suffer a meeting like this, move on. You are unlikely to reach an agreement anytime soon.

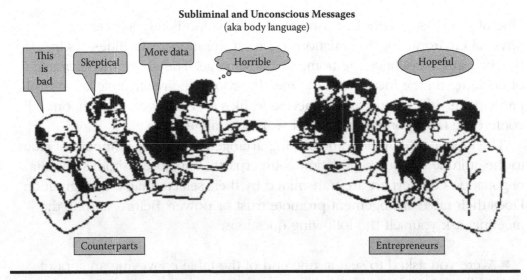

Figure 8.11   Subliminal—What body language are you "seeing" from your negotiating counterparts?

8.7.4 *Power Negotiator's TIR*

"NO is the beginning of negotiations."

Time is money. Few people are more aware of this than are entrepreneurs seeking seed investments.

In negotiations, it simply means that the more time passes, the weaker the entrepreneur's bargaining position becomes. Moreover, investors know this very well.

Power negotiators have three quills in their holster: (1) time, (2) information, and (3) reward power. The entrepreneur must enter negotiations with the full expectation that there will be areas of legitimate differences, and these differences will take time to overcome. You must make time your ally not your adversary.

The entrepreneur brings to the negotiating table a wealth of valuable information. That information takes the form of market knowledge, technical expertise, and competitive intelligence highly sought by potential investors. This is a great source of bargaining power.

The entrepreneur can reward the investor with exceptional returns not available elsewhere. Together these three sources of negotiator's power can bring about a mutually desired outcome, as shown in Figure 8.12.

8.8 Managing Negotiation Conflict

"If you are going through hell, keep on going." —Winston Churchill

Negotiations involve conflict among interdependent entities. Conflict occurs during negotiations over substantive issues. Conflicts can be divided into

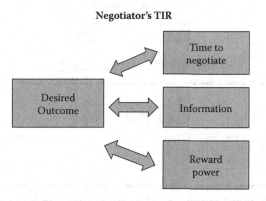

Figure 8.12   Negotiator's TIR—Your sources of negotiating power.

three parts: substantive, emotional, and intergroup. Substantive conflict is seen when there is a fundamental disagreement over goals, while emotional disagreements arise when there is anger, mistrust, fear, and resentment. Intergroup conflict occurs among members of competing teams, especially when groups differ substantially in status, power, and influence.

Conflicts can be overcome with creative solutions and reconsideration of previous positions. Conflicts can be resolved by focusing on objectives and mutual interests. Seek acceptable rather than optimal solutions while ensuring mutual satisfaction.

For example, the parties may have a major conflict over the price of something. The conflict over price threatens to derail the entire agreement. However, price does not need to be a precise number of dollars. Price can have a range, depending on when it is paid, meeting milestones, achieving new specifications, estimated market penetration, etc. Thus, price can have a "settlement range" as shown in Figure 8.13.

The important point is this: Facts are past-oriented; negotiations are future-oriented. Future-oriented settlements can be altered as new information arrives. Ranges can be very useful in overcoming conflicts when they arise.

8.9 Secrets of Power Presentations

"You must be more than a mere messenger."

You will be better believed if your negotiation counterpart considers you an expert and one that can be trusted. Your technical expertise will bring

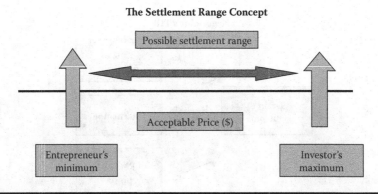

Figure 8.13 The settlement range—Settlement ranges can be used to overcome negotiating conflicts.

you credibility. Your credibility will be based on your intimate knowledge of products, markets, regulations, and future industry directions. Emphasize your experience, accomplishments, and the special qualifications of your team members.

Chester L. Karras,[22] author of *The Negotiating Game* and *Give and Take*, famously states, "In business as in life—you don't get what you deserve, you get what you negotiate." He also admonishes, "Perception is reality." Moreover, everything that happens during negotiations is more perception than reality.

As the chief negotiator for your company, you will create reality by remembering the following truism: The most memorable presentations have three sections: a beginning, middle, and an end. Therefore, follow this dictum: "Tell them what you are going to tell them; tell them and then tell them what you have told them."

■ Disclose a "roadmap" or "plan" of your presentation before launching into your discussion—"Tell them what you are going to tell them."
■ Clearly express the need at the beginning—"This is the market pain."
■ It is more effective to present both sides of an issue up front.
■ When discussing advantages/disadvantages of an issue, unveil your position last.
■ Listeners will remember the beginning and end of a presentation more than the middle.
■ Listeners will remember the end better than the beginning.
■ Listener will only remember bits and pieces of the middle—"Everything they said is in a fog," "I cannot remember precisely how it was said."
■ Repetition leads to better memorization and acceptance.
■ Conclusions should be clearly stated—"This is how we intend to solve this market pain."
■ Summarize your findings—"Tell them what you just told them."

8.10 Secrets of Closing the Deal

"If you can't get a meal, get a sandwich."

Entrepreneurs need to understand that most concessions will occur at the deadline. Alternatively, using Pareto's Law, 80% of the concessions will occur during the last 20% of negotiation time. The crucial question for the entrepreneur is do you have the "staying power" necessary to wait for a fair outcome?

Herb Cohen, author of the bestseller *You Can Negotiate Anything,* states that "the more energy is expended in a goal, the more desirable that goal becomes" (for both parties).[23]

The following summarizes important closing criteria:

- It will take twice as long as you expected to close the deal.
- It will cost you twice as much as you expected to close the deal.
- Agree on common issues first, and then move on to the more difficult ones.
- Focus your counterpart on your innovation (differentiation, positioning).
- Present proposals on a rational basis, using statistics and facts.
- Do not allow the negotiations to be reduced to a single issue (these are zero-sum).
- Bring down the curtain slowly by slowing your concessions as time passes.
- Provide as many options, choices, and approaches as possible.
- In negotiations, style supersedes substance (players count more than the play).[24]

Figure 8.14 visually presents the negotiation process as a funnel. The top of the funnel casts a wide net to the business world called **information sharing**, where the entrepreneur looks at many potential investors and exchanges information with all comers. Once an investor shows interest in reaching a deal, the negotiations move to the **problem census** phase, where deal limits become better defined by both parties.

If the parties are still interested, the most difficult phase starts, which is labeled **problem solving**. Problem solving or conflict resolution allows the parties to proceed or ends in an impasse, stalemate, or deadlock. This phase is the most sensitive part of any negotiation, requiring the most skill and sensitivity by both parties. The majority of your time, effort, and energy will be spent here.

If technical/market problems can be overcome, the parties can move to the stem of the funnel. It is at this **deal terms** phase when the experts (i.e., accountants, lawyers, financial advisors) come into play. The entrepreneur should be particularly concerned with the counterpart's lawyers. Lawyers are the world's worst deal-killers. By concentrating on their client's interests, lawyers overlook the principle that business negotiations

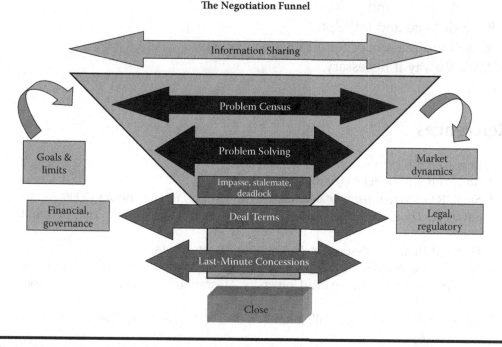

Figure 8.14   The negotiation funnel—Closing the deal in visual form.

are based on mutual compromise. Lawyers seem more interested in PRINCIPAL than in PRINCIPLE.

It is almost axiomatic that counterparts will come up with some last-minute demands, just when their negotiating advantage is at its greatest. It is the proverbial "Oh, by the way...." Your counterpart asks for last-minute concessions. It occurs precisely at your point of maximal negotiation vulnerability. More entrepreneurs have lost control of their startups at this point than at any other point during negotiations.

8.11 The Ten Amendments of Negotiations

You have the right to:

1. Negotiate with decision makers only.
2. Be backed by your own experts.
3. Ask for more time.
4. Concede slowly and ask for a return concession.
5. Ask for clarification.
6. Look intimidated.

7. Look and act indecisive.
8. Be disliked and unhelpful.
9. Demand references for all their statements.
10. Walk away if necessary.

References

1. http://en.wikipedia.org/wiki/Negotiation
2. http://en.wikipedia.org/wiki/Bargaining
3. Shell, R.G. *Bargaining for Advantage*. New York: Penguin Books, 2006.
4. Saner, R. *The Expert Negotiator*. The Netherlands: Kluwer Law International, 2000.
5. Harvard Business Essentials, Negotiation, Boston, MA.
6. http://www.pon.harvard.edu/. The Program on Negotiation (PON) is a consortium program of Harvard University, Massachusetts Institute of Technology, and Tufts University.
7. http://highered.mcgraw-hill.com/sites/dl/free/0070979960/894027/lew79960_chapter02.pdf
8. http://www.prenhall.com/behindthebook/0131868667/pdf/CarrellCh04final.pdf
9. Brazeal, G. Against gridlock: the viability of interest-based legislative negotiation. *Harvard Law & Policy Review* (Online), 3, 1, 2009.
10. Fisher, R. and Ury, W. *Getting to Yes. Negotiating Agreement Without Giving In*. New York: Penguin Books, 1981.
11. http://en.wikipedia.org/wiki/Zone_of_possible_agreement
12. Lewicki, R.J., Minton, J., & Saunders, D. Zone of potential agreement. In: *Negotiation*, 3rd ed. Burr Ridge, IL: Irwin-McGraw Hill, 1999.
13. http://www.harveymackay.com/about/harveys-books/excerpts-from-swim-with-the-sharks-without-being-eaten-alive/
14. Shell, G.R. Bargaining styles and negotiation: The Thomas-Kilmann conflict mode instrument in negotiation training. *Negotiation Journal*, 155–174, 2001.
15. http://en.wikipedia.org/wiki/Negotiation
16. Gates, S. *The Negotiation Book*. United Kingdom: John Wiley & Sons, 2011, p. 240.
17. Goldman, A. *Settling For More: Mastering Negotiating Strategies and Techniques*. Washington, D.C.: The Bureau of National Affairs, Inc., 1991, p. 83.
18. Lewicki, R.J., Saunders, D.M., & Minton, J.W. *Essentials of Negotiation*. New York: McGraw-Hill Higher Education, 2001, p. 82.
19. Karrass, C.L. *Give and Take*. New York: Thomas Y. Crowell Publishers, 1974.
20. Dawson, R. *Secrets of Power Negotiating*, 15th ed. Franklin Lakes, NJ: Career Press, 2001.

21. Sander, F.E.A. How to break a stalemate. Harvard Business School, Reprint No. N0406.
22. Karrass, C.L. *In Business as in Life—You Don't Get What You Deserve, You Get What You Negotiate*. Beverly Hills, CA: Stanford St. Press, 1996.
23. Cohen, H. *You Can Negotiate Anything*. Bantam Books, 1980.
24. Cohen, H. *Negotiating the Game*. Harper Audio, 1993.

Chapter 9

Marketing and Sales

9.1 Introduction

Marketing is the process for creating, communicating, delivering, and exchanging company offerings to customers and stakeholders. Marketing is a critical business function for attracting customers.[1] **Selling** is offering to exchange an item of value for a different item. The original item of value being offered may be either tangible or intangible. The seller most often sees the second item, usually money, as being of equal or greater value than others being offered for sale.[2] A **sale** is the act of exchanging a product or service in return for money or other form of compensation.

The marketing and sales functions differ greatly, but have the same end-goals as shown in Table 9.1. Selling is the final stage in marketing. Both marketing and sales share the goal of increasing the desirability and value to the customer. Achieving this goal may involve the sales team using promotional techniques such as advertising, sales promotion, publicity, and public relations, creating new sales channels, or creating new/innovative products, among other things. It includes bringing the potential customer to (1) visit the organization's website for more information, (2) directly contact the organization for more information, or (3) interact with the organization via social media such as Twitter, Facebook, and blogs.[3]

A manufacturer may use a marketing **pull strategy** by focusing its promotional efforts on opinion leaders, celebrities, or end customers to create a demand or "pull." Pharmaceutical companies are now publically informing consumers about the availability of new prescription drugs. The public is

Table 9.1   Differences between Marketing and Sales

Marketing	Sales
Produces what the customer wants	Tries to get the customer to want what the company produces
Focuses on needs, wants, and demands of buyer	Revolves around needs, interests, and timing of the seller
Outward looking	Inward looking
Creates a pull	Creates a push
Involves a set of interrelated activities	End result of the marketing process
Creates satisfaction through relationships	Personal attention and follow-up
Starts before a product/service exists	After product is manufactured or service created

then urged to "see your doctor for a prescription." This is a classical marketing strategy to stimulate demand and create product sales.

Conversely, a manufacturer may use aggressive selling and trade advertising to convince a retailer, wholesaler, or distributor to carry and promote its product. This approach is known as a **push strategy**. The retailer, in turn, uses advertising, point-of-sale displays, shelf space, etc. to convince the customer to purchase the "pushed" product.[4]

These two classical marketing strategies are shown in Figure 9.1.

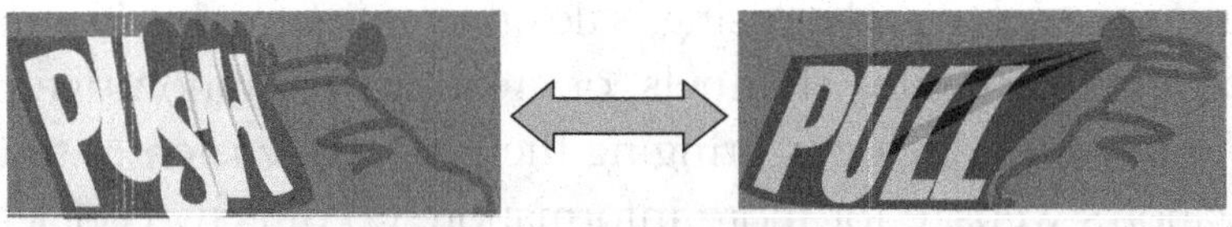

Classical Marketing Strategies

Push Strategy
• Uses the sales force and trade promotions.
• Persuade wholesalers and retailers to carry brands.
• Specialty advertising.
• Cooperative advertising
• Relies on testimonials

Pull Strategy
• Spends on advertising and sales promotion to increase demand.
• Entices customers to try your new product.
• Lures customers away from competitive products
• Lower price; better quality

Figure 9.1   Classical marketing strategies—The two basic marketing strategies of push and pull and their differences.

However, in the final analysis, both marketing and sales strive to attract and retain a critical mass of customers, in order to financially sustain the enterprise. The **critical mass** is a sufficient number of adopters of an innovation in a social system so that the rate of adoption becomes self-sustaining and creates further growth. It is an aspect of the theory of diffusion of innovations, proposed by Everett Rogers in his influential book *Diffusion of Innovations.*[5]

9.2 Selling Your Innovation

"If we listened to the experts, we would be riding faster horses."
—Henry Ford

A startup generally depends on product innovation. An innovation, by definition, is a brand new product/service being offered for the first time. How do entrepreneurs promote their innovation, since most potential customers are unaware of the existence of the product/service?

One of the main functions of marketing is to create an insatiable need. In 2005, how many of us needed a smart phone, or an iPad? Marketing created the need for these products. Prior to the introduction of any of their innovative products, Apple's Steve Jobs directed his marketing department to perform market intelligence, followed by market research. At Apple, the marketing department was charged with the responsibility of coordinating data collection, interpretation of results, and developing a basis for marketing/sales action.

You can promote your innovative products by:

■ Identifying market needs (market pain)
■ Communicating the features/advantages to prospective buyers
■ Making goods/services readily available at convenient times/places
■ Pricing goods/services that reflect costs, competition, and ability/willingness to purchase
■ Providing post-sale service and follow-up to ensure satisfaction

Where in the marketplace does your new product offering fit? Figure 9.2 presents a bird's-eye view of the main categories, according to newness to the company and the market.

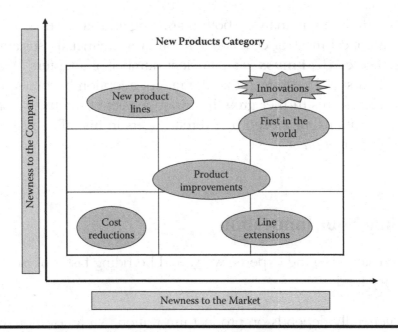

Figure 9.2 New products category—Where does your product fit?

9.2.1 Needs, Wants, and Demands

"Companies can't give job security. Only customers can."
—Jack Welch, CEO, General Electric

Marketing theory divides human necessities into three basic parts. **Needs** comprise some of the most basic and fundamental necessities of life such as food, shelter, protection, good health, etc. These needs are not created by marketing because they already exist in society.

Wants are desires for things that satisfy deeper requests, such as gourmet foods, sports cars, vacations to exotic locales, etc. Advanced societies are continually reshaping wants by societal forces such as schools, families, business corporations, healthcare alternatives, etc. Entrepreneurial companies are very active in satisfying consumer wants with innovative solutions and approaches. Needs are few; wants are many. Remember the old adage— "People don't know what they want, only what they know."

Demands are wants for specific high-quality, high-priced products/ services that deliver superior performance. Entrepreneurial firms shine in this marketing sphere. Most of the innovations carry a hefty price tag, thus satisfying the demands of affluent customers who are willing and able to

Figure 9.3   Marketing program—The six promotional methods available to startups.

buy these offerings. If we use the 80/20 rule, 20% of customers purchase 80% of the demand offerings.

Startups can market and promote their products by utilizing the methods shown in Figure 9.3.

9.3 Analyzing Your Market

"The future isn't what it used to be." —Yogi Berra

Your market is the set of actual and potential buyers of your products/ services. Markets can be divided into four categories: (1) potential, (2) available, (3) qualified, and (4) target. Your **potential** market is the aggregate of all individuals, firms, and organizations in a particular market that have some level of interest in a particular product/service.[6] Your **available** market is where customers are willing and able to purchase the offering. A **qualified** market occurs when prospects have indicated interest by performing some act; that is, downloading specifications, attending a seminar, visiting your booth at a trade show, or calling your company in search of information. Last, your **target** market refers to that select group of prospects that deserve concentrated attention.

Table 9.2   Strategic Orientations

1. Market orientation	2. Sales orientation
Based on the customer's perceived values, needs/wants/demands. Focuses on delivering superior customer relationship.	The sales approach stresses those products that the organization can best produce at the most competitive price.
3. Production orientation	4. Societal orientation
Focuses on the internal manufacturing capability and resources of the firm to manufacture quality products.	A philosophical stance that focuses on preserving or enhancing individual's or society's long-term interest's.

You can decide, up front, your marketing orientation. Your marketing orientation will strongly influence your culture and promotional behavior. As Table 9.2 shows, there are four recognized orientations.

9.3.1 *The Promise of Unmet Needs/Wants*

"Ideas, products, messages and behaviors spread just like viruses do." —Malcolm Gladwell, *The Tipping Point*

Successful startups are those firms that can recognize and respond to unmet needs/demands in the market. Unmet needs/demands are plentiful, such as cure for cancer and heart disease, non-polluting cars, green and sustainable agriculture, desalination of seawater, wireless communications, etc. However, you can analyze your markets all you want, but some parameters will be uncontrollable, such as:

- Demographic trends
- Competitive technologies
- Political/legal environment
- National income

Startups need to avoid marketing myopia. Marketing myopia is management's inability to recognize the scope of its business. To avoid marketing myopia, startups must broadly define their strategic goals, and concentrate on universal customer needs/wants, as depicted in Table 9.3.

Table 9.3 Avoiding Marketing Myopia

Company	Myopic Description	Benefits Description
Apple	We are in the device business.	We are in the wireless business.
JetBlue Airways	We are in the airline business.	We are in the affordable-transportation business.
Morgan Stanley	We are in the brokerage business.	We are in the financial services business.
Revlon	We are in the cosmetics business.	We are in the "hope" business.

9.3.2 Costs and Functions of Marketing

Marketing creates utility (wants/needs/demands satisfying power) through the selling process. The American Marketing Association defines marketing as "An organizational function and set of processes for creating, communicating and delivering value to customers and for managing customer relationships in ways that benefit the organization and its stakeholders." Using this broad definition, we can enumerate the eight costs and functions of marketing, as shown in Table 9.4.

Table 9.4 Costs and Functions of Marketing[7]

1. Buying	Ensuring product availability to meet market demands
2. Selling	Personal contacts, technical support, follow-up, product literature
3. Standardization	Ensuring quality control and quality assurance for all offerings Meeting or exceeding world quality standards
4. Financial assistance	Providing credit, appropriate payment schedules, referrals
5. Storage	Maintaining warehouse facilities to ensure timely deliveries
6. Transportation	Facilitating the delivery of product from production sites to purchasers
7. Market information	Collecting, analyzing, and acting on market dynamics
8. Risk taking	Managing market uncertainty and future customer demands

9.4 Pioneering StartUps

"There are known knowns. There are known unknowns. But there are also unknown unknowns." —Donald Rumsfeld

Introducing a product that is new to both the firm and the market requires the greatest expenditure of both effort and resources, thus representing the greatest uncertainty and risk. The marketing challenge is twofold: (1) communicating the product/service availability and (2) persuading the early adopters.

Figure 9.4 is the classical adoption cycle that accompanies any innovation into the market. The entrepreneur's marketing challenge is to quickly move from the innovator/early adopter cycle to the majority of customers.

As the figure shows, the early market consists of the **innovators** or **techies** who are eager to be at the forefront of any technology, and make a name for themselves if it is proven successful. For the **visionaries,** the technology represents a change agent, and they are the champions of change and set the standards. For the **pragmatists**, improving technology is their priority—they want "evolution not revolution." The **conservatives** wait until a platform or application becomes the *de facto* industry standard, and are price-sensitive and very demanding. The **laggards**

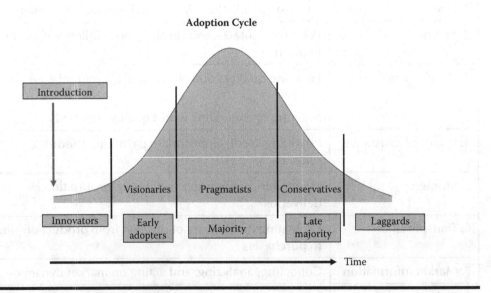

Figure 9.4   Adoption cycle—The classical adoption cycle that accompanies innovative entries.

(risk allergic) wait for turnkey applications with guaranteed results and no surprises.

9.4.1 Crossing the "Chasm"

"If people don't want to come out to the ballpark, who is going to stop them?" —Yogi Berra

For any startup, crossing the chasm, that is, becoming the industry standard and becoming profitable, is its most pressing strategic goal. To accomplish that goal, the firm must persuade the pragmatists to adopt their product/ service. Pragmatists demand a total solution to their problem ("pain"); they want the whole product. The whole product is defined as the minimum set of market requirements to compel a radical change and achieve market superiority.

Crossing the chasm is the whole enchilada. The "chasm" concept is visually presented in Figure 9.5.

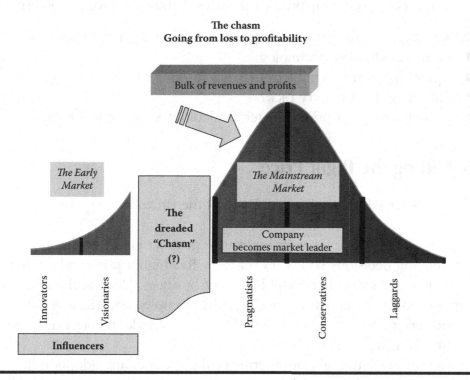

Figure 9.5   The chasm—Becoming the industry standard and reaching profitability.

9.4.2 Your Pioneer Strategy

"Pioneers are the ones with the tomahawks on their backs."

Question: How can a startup aspire to become a market leader in the phase of gigantic competition? Answer: By the marketing concept of a pioneer strategy.

Marketing gurus have long debated the following conundrum: "Is it better to be a pioneer or a follower?" Potential sources of competitive advantage available to pioneers are:

■ First mover advantage
■ First choice of market segmentation
■ Defining the rules of standardization
■ Early distribution advantage
■ Setting prices (optimizing margins and revenues)
■ Creating high barriers to entry

In spite of these early entry advantages, many pioneers fail to capitalize on their potential by abandoning the product, going out of business, or being acquired before their product matures. Followers have the ability to:

■ Exploit any pioneer mistakes (product, marketing, technical)
■ Use more advanced technologies
■ Exploit the pioneer's limited financial resources
■ Skim the market (cherry pick)
■ Compete on lower price, greater availability, superior technical support

9.5 Setting the Right Price

"You don't sell through price. You sell the price."

Setting the "right" price for your innovative product is one of the most crucial marketing decisions you will ever face. Too high a price and it will not sell; too low a price and you will be out of business. Meeting the price of existing products is the easiest pricing goal to implement. However, how do you price an innovative product/service because by definition it does not currently exist in the market?

Pricing must be thought of in terms of the product and adoption life cycle. Keep in mind that pricing is the only part of the marketing mix that produces revenues; all the other elements produce costs.[8] Your selected pricing strategy will:

- Define your product
- Help to segment the market
- Incentivize customer adoption
- Signal your quality intentions to your competition
- Establish the gold standard

Figure 9.6 summarizes the recommended six steps in setting your pricing policy.

9.5.1 Establishing Your Pricing Tactics

"I never said most of the things I said." —Yogi Berra

Your pricing tactics should accurately reflect your strategic goals. As an innovative startup, you should be thinking in terms of value, not just pricing. Value in new product pricing ensures that customers receive fair value-based pricing, while enabling the entrepreneur to reach an industry price equilibrium that provides adequate revenue returns.[9]

Value is the difference between what the customer gains from owning a product and the costs of obtaining the product. **Quality** is the

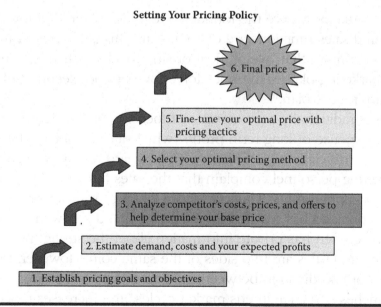

Setting Your Pricing Policy

6. Final price

5. Fine-tune your optimal price with pricing tactics

4. Select your optimal pricing method

3. Analyze competitor's costs, prices, and offers to help determine your base price

2. Estimate demand, costs and your expected profits

1. Establish pricing goals and objectives

Figure 9.6   Setting your pricing policy—Your most important marketing decision.

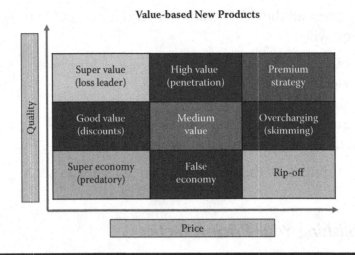

Figure 9.7   Value-based products—The price-quality continuum.

characteristics of a product/service that satisfy stated or implied customer needs. Value-based new product offerings can best be seen in terms of a price-quality continuum, as shown in Figure 9.7.

9.5.2 Tensions between Marketing and Sales

"If you come to a fork in the road, take it." —Yogi Berra

In a startup situation where the founders are critically looking for sales, the marketing and sales functions are often at war. This is in spite of the fact that everyone knows that "we all float or sink together." If sales are disappointing, marketing blames the sales force for its poor execution of an otherwise brilliant rollout plan.[10]

Marketing traditionally accuses sales of focusing on pricing and short-term sales at the expense of long-term profits, while sales complains bitterly that prices are set artificially high without proper regard to current market conditions. Marketing personnel complain that the sales force is focused on closing deals, forsaking profits. The sales force lobbies for lower prices because the product demand has reached its elastic limit, and it is well accepted that the sales force often has private information about the strength of demand.[11]

Marketing and sales are two sides of the same coin. However, there is generally poor coordination between the two groups, which only raises market launches costs, lengthens market cycles, and increases rancor and internal disharmony. The founder must establish an unambiguous "revenue

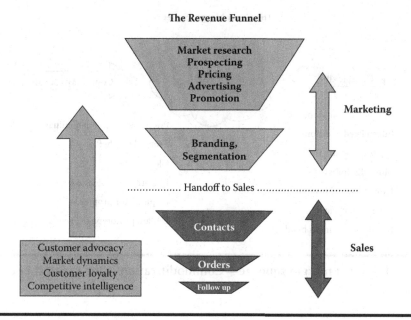

Figure 9.8   The revenue funnel—Responsibilities of marketing and sales.

funnel" that clearly defines the responsibilities of each group and what are management's realistic expectations for revenues and sustained sales growth, as shown in Figure 9.8.

9.6 The Complex Sale

"Focus on success, not failure avoidance."

A **complex sale** is a type of selling where a number of people must give their approval or input before the buying decision can be made. Complex sales are primarily focused on business-to-business and business-to-government transactions, and can range from a few weeks to years.[12]

Most innovative startups will face the challenge of the complex sale. The current business environment is characterized by four interrelated phenomena: (1) escalating customer requirements with increased complexity, (2) rapid commoditization leading to price erosion, (3) relentless competitive forces, and (4) need to respond within a tight window of opportunity. Commoditization is the pressure exerted by the customer to equalize the differences between suppliers, thus reducing their decision-making to the lowest common denominator: the selling price. This leads to the great margin squeeze as depicted in Figure 9.9.

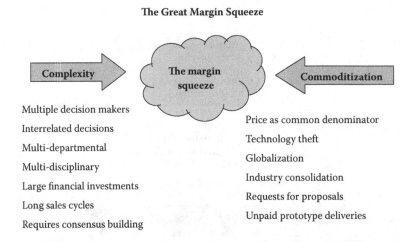

The Great Margin Squeeze

Complexity → The margin squeeze ← Commoditization

Multiple decision makers
Interrelated decisions
Multi-departmental
Multi-disciplinary
Large financial investments
Long sales cycles
Requires consensus building

Price as common denominator
Technology theft
Globalization
Industry consolidation
Requests for proposals
Unpaid prototype deliveries

Figure 9.9   The great margin squeeze—Commoditization is the greatest enemy of profit margins.

The length of time that an innovation enjoys the advantage of being first in the market is getting shorter and shorter. For a startup, the key to success is to *differentiate* your innovative offering. Differentiation allows for more profitable, preemptive, and effective product introductions.

9.7 Market Segmentation

Market segmentation is a marketing strategy that involves dividing a broad target market into subsets of consumers who have common needs, characteristics, and behavior patterns. The firm can then design and implement strategies to target customer needs and desires using media channels and other touch-points that best allow reaching them.[13]

The market segmentation fundamental thesis is that it achieves a competitive advantage by (1) identifying demand segments, (2) targeting specific customer segments, and (3) developing marketing "mixes" for each targeted market segment.[14] As a result, market segmentation enables companies to target different categories of consumers who perceive the full value of certain products and services differently from one another. Generally, four criteria can be used to identify different market segments, as follows:

1. Homogeneity (common needs within one segment)
2. Distinction (unique from other groups in the category)

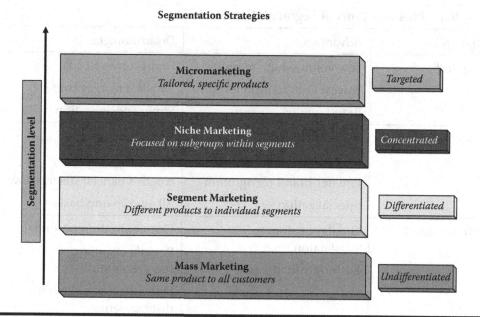

Figure 9.10 Segmentation strategies—Achieving optimal value through four market segmentation strategies.

3. Heterogeneity (individuals with diverse product needs)
4. Reaction (similar response to market forces)[15]

A market segment consists of individuals, groups, or organizations with one or more characteristics that cause them to have relatively similar product needs. Figure 9.10 presents the recognized market segmentation strategies.

Table 9.5 discusses the advantages and disadvantages of the different segmentation strategies.

9.8 Positioning: Your Place in the Sun

"Not what you can do to the product, what you can do to the mind." —Jack Trout

In the marketing sphere, **differentiation** is the process of creating an intellectual image to clearly distinguish the company's products from the competition.[16] The key to differentiation is positioning. **Positioning** is the process of creating a desired image of your company and its products in the minds of potential customers and stakeholders. Positioning is a marketing concept that was first introduced by Jack Trout[17] and then popularized

Table 9.5 Pros and Cons of Segmentation Strategies

Strategy	Advantages	Disadvantages
Targeted	Low volume/high margins Lowest competition level Brand equity	Highest overall cost Smallest market share Super specialization
Concentrated	Better designs Compete with larger firms Greater brand recognition Specialization	Smaller markets Commoditization Needs financial staying power All eggs in one basket
Differentiated	Greater customer satisfaction Better profit margins	Cannibalization potential Rapidly changing requirements Difficulty in maintaining distinctiveness
Undifferentiated	Lowest marketing costs Savings on production Lowest development costs High volume/low margins	Unimaginative offering Susceptible to lower cost competition High distribution costs Globalization

by Al Ries and Jack Trout in their bestseller book *Positioning—The Battle for Your Mind*.[18]

Positioning is the great antidote to commoditization in your market. The centrality of positioning is shown graphically in Figure 9.11.

9.8.1 Famous Positioning Statements

The **positioning statement** is a subset of an optimized value proposition for marketing communications purposes. It identifies the target audience, the product and its category, and a specific benefit, and is differentiable from the nearest competitive alternative. It is a *non-ambiguous statement* that becomes the messaging cornerstone of an integrated marketing campaign, as shown here:

■ Target Audience: The attitudinal and demographic description of the core prospects. The group of customers that most closely represents the brand's most passionate and enthusiastic users.

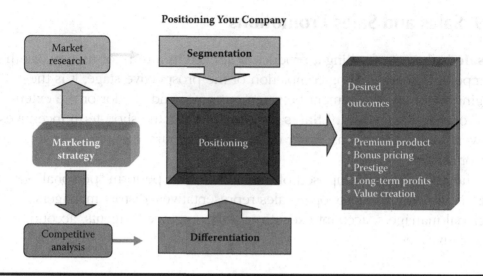

Figure 9.11   Positioning your company—Distinguishing your products from the competition through positioning.

- Frame of Reference: The specific category in which the brand competes. Provides brand relevance to the customer.
- Differentiation: The most compelling and motivating benefit *relative* to the competition.
- Credibility: The most convincing proof that the brand will deliver its promises.[19]

9.8.2 Template for a Positioning Statement

"For *(target audience)*, is the *(frame of reference)* that delivers *(differentiation)* because only *(brand name)* is *(credible)*". As inspiration, read through these positioning statements from large and small companies:

- Mercedes-Benz: "Engineered like no other car in the world"
- BMW: "The ultimate driving machine"
- Southwest Airlines: "The short-haul, no-frills, and low-priced airline"
- Avis: "We are only Number 2, but we try harder"
- Famous Footwear: "The value shoestore for families"
- Miller Lite: "The only beer with superior taste and low caloric content"
- Lexus: "The relentless pursuit of perfection"
- Beck's: "The only German word you need to know"

9.9 Sales and Sales Promotions

A **sale** is the act of selling a product or service in return for money or other compensation.[20] Signaling completion of the prospective stage, it is the beginning of an engagement between customer and vendor or the extension of that engagement.[21] **Sales promotions** are the short-term incentives (lower price, better terms, faster delivery) to encourage purchase of a product or service.

The sales field is composed of individuals who perform "personal" selling. Examples are salespeople, sales representatives, district managers, regional managers, account executives, sales engineers, agents, account representatives, etc.

9.9.1 Your Sales Force

> "Set the sale goals, then empower your sales force to accomplish those goals."

Salespeople include order takers (behind a counter or by web/telephone) or interpersonal (face-to-face communications or teleconferencing). They may be home-based or out in the field. Keep in mind that salespeople represent the company to customers and, conversely, represent customer interests to the company.

The sales force is managed by the methods shown in Figure 9.12.

9.9.2 Sales Force Compensation

The purpose of the sales force compensation metric is to determine the mix of salary, bonus, and commission that will maximize sales generated by the sales force. When designing a compensation plan for a sales force, managers face four key considerations: level of pay, mix between salary and incentive, measures of performance, and performance-payout relationships.

The level of pay, or compensation, is the aggregate amount that a company plans to pay a salesperson over the course of a given year. This can be viewed as a range because its total will vary with bonuses or commissions.[22] Surveys consistently show that 95% of sales force compensation schemes had a combination of quotas and commissions.[23] Compensation packages can be complex and affected by multiple forces, as seen in Table 9.6.

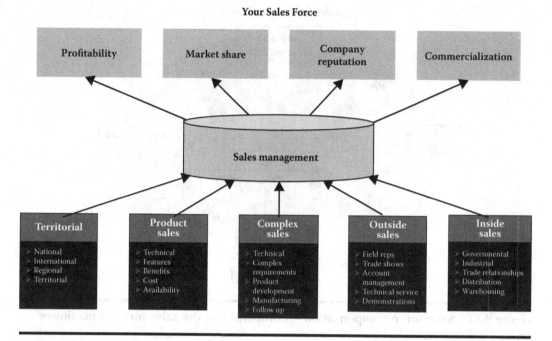

Figure 9.12 Your sales force—Sales force management and goals achievement.

Management enjoys considerable freedom in designing compensation systems. The goal is to start with a reasoned forecast for sales and a range for each salesperson's compensation. After these elements are determined, there are many ways to motivate a salesperson. Key formulas in this area include the following:

Total Compensation Package ($) = Salary ($) + Bonus 1 ($) + Bonus 2 ($)

Total Compensation Package ($) = Salary ($) + [Sales ($) × Commission (%)]

The dynamic effects of sales force compensation coupled with corporate sales objectives are visually presented in Figure 9.13. Our goal is to demonstrate how compensation schemes can affect the sales output.

Table 9.6 Balance between Company Policies and Market-Specific Elements

Financial incentives	Salary, commissions, bonuses, promotions, sales contests
Non-financial incentives	Awards, personal recognition, vacation

Figure 9.13   Sales force compensation—Compensating the sales force to maximize sales.

9.9.3 Sales Force Time Management

While salespeople spend their time widely from company to company, most salespeople report their time is typically divided between five major activities as shown in Table 9.7.

To apply the 80/20 rule to sales, only 20% of your time results in actual orders. As a result, salespeople should:

■ Focus your time on your highest-potential customers.
■ Use qualified leads and referrals only.
■ Focus on results, not activities.

Table 9.7   Time-Activities Performed by Salespeople

Weekly Percent Time Spent	Activity
30	Personal selling
25	Phone calls
20	Traveling
17	Administrative activities
8	Service

■ Spend as little time as possible on non-revenue activities.
■ Talk to decision-makers only.

References

1. http://en.wikipedia.org/wiki/Marketing
2. http://en.wikipedia.org/wiki/Selling
3. http://en.wikipedia.org/wiki/Sales
4. Promotional Mix Strategies. http://www.londremarketing.com/documents/Pushpull08102005.ppt
5. Rogers, E.M. *Diffusion of Innovations*. New York: Simon & Schuster, 2003.
6. http://www.businessdictionary.com/definition/potential-market.html
7. Modified after Marketing: Creating Satisfaction through Customer Relationships. Chapter 1. http://www.sba.pdx.edu/faculty/johne/316/Ch1.ppt
8. Kotler, P. *Marketing Management*. Englewood Cliffs, NJ: Simon & Schuster Company, 1994.
9. Bernstein, J., & Macias, D. Engineering new product success: the new product pricing process at Emerson. *Industrial Marketing Management*, 31, 51–64, 2002.
10. Kotler, P., Rackaham, N., & Krishnaswamy, S. Ending the war between sales and marketing. *Harvard Business Review*, reprint R0607E.
11. Simester, D., & Zhang, J. Tension at the marketing–sales interface: why do sales people spend so much time lobbying for low prices? MIT Sloan School of Management. http://research.chicagobooth.edu/marketing/qme/docs/Session8_SimesterEtAl.pdf.
12. Thull, J. *Mastering the Complex Sale*. New York: John Wiley & Sons, 2000.
13. http://en.wikipedia.org/wiki/Market_segmentation
14. Hunt, S.D., & Arnett, D.B. Market segmentation strategy, competitive advantage, and public policy; grounding segmentation strategy in resource-advantage theory. *Australasian Marketing Journal*, 12(1), 7–25, 2004. http://wwwdocs.fce.unsw.edu.au/marketing/12_01_hunt.pdf
15. http://www.investopedia.com/terms/m/marketsegmentation.asp
16. Levi, K. Differentiate or diminish: the art and necessity of business positioning, 9, 2007. www.winningmessage.com
17. Trout, J. "Positioning" is a game people play in today's me-too market place. *Industrial Marketing*, 54(6), 51–55, 1969.
18. Ries, A., & Trout, J. *Positioning: The Battle for Your Mind*. New York: Warner Books–McGraw-Hill, 1981.
19. http://www.brandeo.com/positioning%20statement
20. http://dictionary.reference.com/browse/sales
21. http://en.wikipedia.org/wiki/Sales

22. Farris, P.W., Bendl, N.T., Pfeifer, P.E., & Reibstein, D.J. *Marketing Metrics: The Definitive Guide to Measuring Marketing Performance.* Upper Saddle River, NJ: Pearson Education, 2010.
23. Nair, H., & Misra, S. A Structural Model of Sales-Force Compensation Dynamics: Estimation and Field Implementation. http://faculty-gsb.stanford.edu/nair/documents/MisraNair_StrucuralSalesforceEstimationFieldImplementation.pdf

Chapter 10

Intellectual Property

10.1 Introduction

Intellectual property (**IP**) is a legal concept that refers to creations of the mind for which exclusive rights are recognized.[1] In its broadest sense, intellectual property means the legal rights that result for intellectual activity in the industrial, scientific, literary, and artistic fields, and protection against unfair competition.[2]

On July 31, 1790, Samuel Hopkins was issued the first patent for a process of utilizing potash, an ingredient used in fertilizer, for soap making. The patent was signed by President George Washington, by Secretary of State Thomas Jefferson, as well as the Secretary of War and the Attorney General. Hopkins was born in Vermont, but was living in Philadelphia, PA, when the patent was granted and issued as shown in Figure 10.1.

Under the international intellectual property law concept, owners are granted certain exclusive time-limited rights (20 years from the filing date) to a variety of intangible assets, such as musical, literary, and artistic works; technical discoveries and inventions; plus words, phrases, expressions, symbols, designs, commercial names, and designations. It is your bank account of ideas.

Familiarity with the concept of intellectual property can give you an edge in today's competitive environment. Legally recognized intellectual property comes in several forms, as summarized in Table 10.1.

A **patent** grants an inventor exclusive rights to make, use, sell, and import an invention for a limited period of time (on June 8, 1995, the new

Figure 10.1   First patent issued—Signed by none other than President George Washington.

term took effect in the U.S.), in exchange for the public disclosure of the invention and its practical application. An invention is a solution to a specific technological problem, which may be a product or a process.

You cannot patent naturally occurring products in nature, scientific principles, laws of nature, mental processes, and mathematical formulas.

In the past, there was a great deal of variation of the term of protection afforded by patents in different countries. The members of the World Trade Organization (formerly GATT) have now harmonized recognition of technology patents for a 20-year period that begins with the priority date.

An **industrial design right** protects the visual design of objects that are not purely utilitarian. An industrial design consists of the creation of a shape, configuration, or composition of pattern or color, or combination of pattern and color in three-dimensional form containing aesthetic value. An industrial design can be a two- or three-dimensional pattern used to produce a product, industrial commodity, or handicraft.

A **copyright** gives the creator of original work exclusive rights, usually for a limited time. Copyright may apply to a wide range of creative, intellectual, or artistic forms, or "works." Copyright does not cover ideas and information themselves, only the form or manner in which they are expressed.

Table 10.1   Intellectual Property

Patents	Provide rights for up to 20 years for inventions. A patent does not grant an absolute right to make or sell the invention.
Utility patents	Useful processes, machines, articles of manufacture, and compositions of matter. Examples: fiber optics, computer hardware, medications.
Design patents	New, original, and ornamental designs for articles of manufacture. The look of an athletic shoe, look-and-feel of software, a bicycle helmet, and the Star Wars characters are all protected by design patents.
Plant patents	Invented or discovered, asexually reproduced plant varieties. Hybrid tea roses, Silver Queen corn, Better Boy tomatoes, Russet potatoes, and Madelia onions are all types of plant patents.
Trademarks	Words, names, symbols, sounds, or colors that distinguish goods and services. Trademarks, unlike patents, can be renewed forever as long as they are being used in business. The roar of the MGM lion, the pink of the Owens-Corning insulation, and the shape of a Coca-Cola bottle are familiar trademarks.
Copyrights	Works of authorship, such as writings, music, and works of art that have been tangibly expressed. The Library of Congress registers copyrights created by an individual, which last the life of the author plus 70 years. For a work created for an employer, the copyright lasts the shorter of 95 years from publication or 120 years from the date of creation.
Trade Secrets	Information that companies keep secret to give them an advantage over their competitors. The formula for Coca-Cola is one of the most famous trade secrets.
Trade dress	A legal term of art that generally refers to characteristics of the visual appearance of a product or its packaging (or even the design of a building) that signify the source of the product to consumers.

A **trademark** is a recognizable sign, design, or expression that identifies products or services of a particular source from those of others. Ford, Lexus, Apple, and Microsoft are examples of trademarks.

A **service mark** is any word, symbol, or phrase that identifies services rendered by a company and distinguishes them from others; for example, "Fly the Friendly Skies" by United Airlines.

Trade dress is a legal term of art that generally refers to characteristics of the visual appearance of a product or its packaging (or even the design of a building) that signify the source of the product to consumers. A trade dress is universally conferred to pharmaceutical products to properly identify the drug and its manufacturer, and as a competitive differentiator.

A **trade secret** is a formula, practice, process, design, instrument, pattern, or compilation of information that is not generally known or reasonably ascertainable, by which a business can obtain an economic advantage over competitors or customers. Following are some techniques for creating and maintaining trade secrets:

■ Keeping private and confidential documentation
■ Restricting access to all forms of confidential information
■ Establishing a security system for maintaining secrecy
■ Controlling visitors' access to documents or facilities
■ Requiring written employee secrecy agreements
■ Conduct new hire and exit interviews emphasizing secrecy and confidentiality

10.2 Patentable Inventions

"There are no unrealistic goals; only unrealistic time frames."

You can only patent certain inventions: compositions of matter (chemicals), machines/apparatuses, devices, articles of manufacture, processes/methods, ornamental designs and botanical plants, genetically engineered organisms, genetic engineering methods, etc. Table 10.2 lists some of the principles behind intellectual property law.

Table 10.2 Rules of the Game

An invention must be:	Cannot patent an invention if it is:
• New (novelty, enabling disclosure)	• Known or used by others
• Useful (practical)	• Previously described
• Non-obvious to persons "skilled in the art"	• In public use for more than one year
• Searched for "prior art"	• A law of nature or physical phenomenon

An invention is **new** if it was not invented by someone else first; it is **useful** if it works; it is **non-obvious** if it was unexpected and unanticipated. For example, the scientists that created the Post-it® adhesive were actually trying to develop an adhesive that would bond forever. It was a big surprise for them when the non-bonding adhesive could be used repeatedly without leaving a sticky residue.

10.2.1 Who Benefits From Intellectual Property Rights?

What can monopolies such as patents, trademarks, or copyrights do for society? A monopoly, rightfully obtained, gives the owner the right to exclude others from making, using, or selling the invention, or using substantially similar "expressions." For example, a patent gives a benefit to an inventor and a benefit to the public, as shown next:

- The patent gives the public a set of detailed instructions that explains how the patent works. Thus, the inventor is teaching others by contributing to the promotion of national and societal progress.
- Anyone is free to use these techniques as inspiration, reference, or to make new contributions, as long as the results do not infringe the patent while it is in force.

A patent is a bargain between society and an inventor wherein the inventor discloses all inventions to the public in exchange for a time-limited monopoly. This ensures that society will be able to enjoy the full benefit of the invention following the expiration date of the patent. (A U.S. patent expires as of noon on the expiry date.)

10.3 Understanding Patents

"Everything that can be invented has been invented." —Charles H. Duell, Commissioner, U.S. Office of Patents, 1899

Patents are highly stylized documents. In contrast to scientific or technical papers, which presume background knowledge by the reader, a patent must stand on its own. Prior knowledge is not assumed, other than normal reading and comprehension skills in the art addressed by the patent. Each patent is an individual presentation of the problem addressed, current techniques that fall short of solving the problem, the many different ways attempted by inventors over time to overcome the difficulty, and finally the proposed solution to the problem by the inventor.[3]

However, the Patent Office examiner does not edit the terminology used by the applicants. A well-established judicial principle is that "the inventor is his own lexicographer." Terms, conditions, or characterizations are taken to have the meaning expressed and defined by the inventor, unless they are misleading or depart completely from accepted practice.

10.3.1 Contents of a Patent

"He hits from both sides of the plate. He's amphibious." —Yogi Berra

Patent documents nearly always contain certain common elements, in the following order:

Subject	Statement of the technology field
Abstract	Concise summary of the enabling disclosure
Problem to be solved	Prior art; background information
Objects	Benefits provided by the discovery
Definitions	Technical exposition of solution to the problem
	Detailed elaboration of all technical aspects of the invention
Utility	Unambiguous description of the practical usefulness of the invention
Examples	Working examples that, if followed, produce the intended results
Claims	Legal description of what is being granted as rights

The first page of a recent U.S. patent is shown in Figure 10.2.

10.3.2 Claims: The Heart of a Patent

From a business perspective, **patent claims** are the most valuable part of a patent. Patent claims are the legal basis for your "legal monopoly." Claims form a protective boundary line around your patent that lets others know

(12) **United States Patent**
Szycher

(10) **Patent No.:** **US 8,367,094 B2**
(45) **Date of Patent:** **Feb. 5, 2013**

(54) **ANTIMICROBIAL MATERIAL AND METHOD FOR MAKING THE SAME**

(76) Inventor: **Michael Szycher**, Lynnfield, MA (US)

(*) Notice: Subject to any disclaimer, the term of this patent is extended or adjusted under 35 U.S.C. 154(b) by 0 days.

(21) Appl. No.: **13/402,336**

(22) Filed: **Feb. 22, 2012**

(65) **Prior Publication Data**

US 2012/0150095 A1   Jun. 14, 2012

Related U.S. Application Data

(60) Continuation-in-part of application No. 13/178,002, filed on Jul. 7, 2011, now Pat. No. 8,173,151, which is a division of application No. 12/347,703, filed on Dec. 31, 2008, now Pat. No. 7,998,498.

(60) Provisional application No. 61/022,609, filed on Jun. 22, 2008.

(51) **Int. Cl.**
A61F 2/00 (2006.01)

(52) **U.S. Cl.** 424/423

(58) **Field of Classification Search** .... None
See application file for complete search history.

(56) **References Cited**

U.S. PATENT DOCUMENTS

2,943,627	A	4/1960	Howell
4,817,594	A	4/1989	Juhasz
4,895,154	A	1/1990	Bartelt et al.
4,982,742	A	1/1991	Claude
5,205,297	A	4/1993	Montecalvo et al.
5,320,598	A	6/1994	Haak et al.
5,395,398	A	3/1995	Rogozinski
5,445,606	A	8/1995	Haak et al.
6,032,077	A	2/2000	Pomeranz et al.
6,051,748	A	4/2000	Auguste et al.
6,280,434	B1	8/2001	Kinoshita et al.
6,365,220	B1	4/2002	Burrell et al.

6,411,853	B1	6/2002	Millot et al.	
6,716,895	B1 *	4/2004	Terry	523/122
6,821,936	B2	11/2004	Green et al.	
6,865,859	B2	3/2005	Trogolo et al.	
2004/0087877	A1	5/2004	Besz et al.	
2005/0064005	A1	3/2005	Dinh et al.	
2005/0271698	A1	12/2005	Bucay-Couto et al.	
2006/0035039	A1 *	2/2006	Ylitalo et al.	428/32.22
2006/0121078	A1	6/2006	Trogolo et al.	

(Continued)

FOREIGN PATENT DOCUMENTS

EP	0367320 A1	5/1990
EP	0504715 A2	9/1992

(Continued)

OTHER PUBLICATIONS

Lubrizol Technical Data Sheet "Carbothane® B20 TPU Series" (p. 1, Sep. 2011).*

(Continued)

Primary Examiner — Bethany Barham
(74) *Attorney, Agent, or Firm* — William A. Loginov; Loginov & Associates, PLLC

(57)   **ABSTRACT**

This invention provides a modified catheter/indwelling device biomaterial that provides both immediate, and long-term microbiocidal effects on otherwise antibiotic-resistant strains of microorganisms. The material, which exhibits good mechanical performance characteristics for medical devices, is composed of a hydrophobic polyurethane (PU), a hydrophilic polyethylene vinyl acetate (PEVA) as an option, a soluble silver salt and a sparsely-soluble silver salt. The hydrophobic polyurethane provides the good physical properties, the PEVA the hydrophilicity necessary to allow some water ingress into the catheter, the soluble silver salt for an immediate burst effect, and the sparsely-soluble silver salt for sustained-release over many months postimplantation. Alternatively, the sparsely soluble silver salt can be silver iodate and is combined with silver sulfadiazine. Chlorhexidine can also be included in the material.

20 Claims, 4 Drawing Sheets

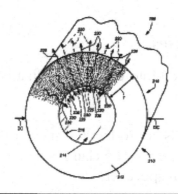

Figure 10.2   First page of a recent patent—A patent follows common elements.

when they may be infringing on your rights. The limits of this line are defined by the words and phrasing of your claims.

Claims are usually in the form of a series of numbered expressions or, more precisely, noun phrases following the description of the invention in a patent or patent application, and define, in technical terms, the extent of the protection conferred by a patent or by a patent application. They are of the utmost importance during both prosecution and litigation.

Ideally, each claim should have only one meaning, which can be either broad or narrow, but not both at the same time. In general, a narrow claim specifies more details than a broader claim. Having many claims, where each one is a different scope, allows you to have legal title to several aspects of your invention. Claims are the legal description of the exclusive rights granted by the government to the inventor. There are two basic types of claims: **independent claims**, which stand on their own, and **dependent claims**, which refer to an earlier cited claim or several claims and generally express particular embodiments as fallback positions. The expressions "in one embodiment," "in a preferred embodiment," "in a particular embodiment," "in an advantageous embodiment," or the like often appear in the description of patent applications and are used to introduce a particular implementation or method of carrying out the invention. These embodiments usually correspond to a dependent claim or could form the basis of a dependent claim. Each dependent claim is, by law, more narrowly defined than the independent claim upon which it depends.

10.3.3 Claims Categories

Claims can also be classified into **categories**, that is, in terms of what they *claim*. A claim can refer to:

- a physical entity, that is, a product (or material) or an apparatus (or device, system, article, etc.). The claim is then called "product claim" or "apparatus claim," respectively
- an activity, that is, a process (or method) or a use. The claim is then called "process claim" (or method claim) or "use claim," respectively

Figure 10.3 shows the claims of a recently issued U.S. patent. Note the independent claim (#1) and all the other dependent claims.

US 8,367,094 B2

15

with differing antimicrobial characteristics using co-extrusion is described, any acceptable manufacturing and/or assembly technique can be employed in alternate embodiments. For example different portions/layers can be adhered, fastened, co-molded, welded together, interlocked, force-fitted, or otherwise joined in alternate embodiments. Note, as used herein the term "composition" in connection with a portion or layer of the device shall refer to a predetermined mixture at least some of PU, PEVA, soluble ionic silver salt and sparsely ionic silver salt (and/or AgSD, or an equivalent thereof). Some or all of these components (and other additional components as desired) can be provided to the material of each portion or layer.

The foregoing has been a detailed description of illustrative embodiments of the invention. Various modifications and additions can be made without departing from the spirit and scope if this invention. Each of the various embodiments described above may be combined with other described embodiments in order to provide multiple features. Furthermore, while the foregoing describes a number of separate embodiments of the apparatus and method of the present invention, what has been described herein is merely illustrative of the application of the principles of the present invention. For example, the compound of this invention can be formed by a variety of mixing techniques using a variety of solvents. In addition, a combination of two or more soluble silver salts and/or two or more sparsely soluble silver salts can be combined in a material so the material can benefit from differing performance characteristics of each of the multiplicity of salts. Also, a finished device constructed from the illustrative material can be constructed by any acceptable mechanism, including, but not limited to, extrusion, injection molding, blow molding and the like. It should also be recognized by those of ordinary skill that the compound described herein can be applied to any acceptable device or shape in which anti-microbial properties are desired, including, but not limited to implantable device, artificial organs, surgical instruments and consumer products. Accordingly, this description is meant to be taken only by way of example, and not to otherwise limit the scope of this invention.

What is claimed is:

1. An antimicrobial material comprising an intimate mixture of:
(a) a slightly hydrophilic polyurethane (PU);
(b) a sparsely soluble ionic silver salt; and
(c) less soluble silver sulfadiazine;
wherein elements (a), (b) and (c) are combined in one operation using a one-stage melt process to form a single unitary structure.

2. The antimicrobial material as set forth in claim 1 wherein the sparsely soluble ionic silver salt comprises silver iodate.

3. The antimicrobial compound as set forth in claim 2 further comprising chlorhexidine.

16

4. The antimicrobial material as set forth in claim 2 further comprising a hydrophilic polymer in a range of between 0.2% and 20% by weight of the material.

5. The antimicrobial material as set forth in claim 2 wherein the sparsely soluble ionic silver salt is between 0.1 and 10% by weight of the material.

6. The antimicrobial material as set forth in claim 2 wherein the sparsely soluble ionic silver salt is between 0.1% and 15% by weight of the material.

7. The antimicrobial material as set forth in claim 1 wherein the sparsely soluble ionic silver compound is silver iodate.

8. The antimicrobial compound as set forth in claim 7 further comprising chlorhexidine.

9. A medical device having a wall constructed and arranged to contact internal tissue and fluids of a patient, the wall including the antimicrobial material as set forth in claim 8.

10. A medical device having a wall constructed and arranged to contact internal tissue and fluids of a patient, the wall including the antimicrobial material as set forth in claim 1.

11. The medical device as set forth in claim 10 wherein the wall includes (a) a first portion having a first composition and (b) a second portion having a second composition.

12. The medical device as set forth in claim 11 wherein the first portion includes at least one of the sparsely soluble ionic silver salt in a first concentration and the sparsely soluble ionic silver salt in a first concentration and (b) a second portion having at least one of the sparsely soluble ionic silver salt in a second concentration, different from the first concentration and the sparsely soluble ionic silver salt in a second concentration, different from the first concentration.

13. The medical device as set forth in claim 10 wherein the wall defines an indwelling catheter shaft and each of the first portion and the second portion define concentric layers of the catheter shaft.

14. The medical device as set forth in claim 13 wherein the wall defines a wall of an indwelling catheter.

15. The medical device as set forth in claim 14 wherein the indwelling catheter comprises a Chronic Dialysis Catheter.

16. The medical device as set forth in claim 14 wherein the indwelling catheter comprises a Peripherally Inserted Central Catheter.

17. The medical device as set forth in claim 14 wherein the indwelling catheter comprises a Urinary Catheter.

18. The medical device as set forth in claim 14 wherein the indwelling catheter comprises a Gastrostomy Catheter.

19. The medical device as set forth in claim 14 wherein the indwelling catheter comprises a Cerebral Spinal Fluid (CSF) shunt.

20. The medical device as set forth in claim 10 wherein the wall is constructed as an extrusion.

* * * * *

Figure 10.3 **The claims page—The heart of a patent are the allowed claims.**

10.3.4 Real Estate Analogy

Claims are considered legal property; that is, they can be bought, sold, rented, or allowed to lie fallow. Patent claims can be likened to the description of a parcel of land in real estate deeds as described in Table 10.3.

Table 10.3 Similarities between Patents and Real Estate Deeds[4]

Patent Terminology	Real Estate Terminology
Claim limits	Metes and bounds that locate and define the perimeter of the property
Exclusion	No trespassing sign, building a fence, limited access
Licensing or demanding a royalty	Charging a fee for entering the property Charging rent
Cross-licensing	Providing common access for mutual benefit
Infringement	Trespassing To enter into the property without permission

10.4 Provisional Patents

For the entrepreneur/inventor trying to be as frugal as possible, a **provisional patent application** may be your savior. Starting on June 8, 1995, the U.S. Patent and Trademark Office (USPTO) allows inventors the option of filing a provisional application for **utility (mechanical, electrical, or chemical) patents**. The keyword in provisional patent is "provisional." A provisional patent only gives one year of protection. After that, you must file for a non-provisional patent or abandon your patent.

A provisional patent is a low-cost alternative, a *preliminary* step before filing for a regular patent that gives one additional year of protection or grace—maybe enough time to test market your invention before investing in the full cost of a regular patent. A provisional patent allows filing without any formal patent claims, oath or declaration, or any information disclosure (prior art) statement.

- ■ It provides the means to establish an early effective filing date in a non-provisional patent application (also known as a docket).
- ■ It also allows the term "Patent Pending" to be applied to your invention.

10.4.1 Time Limits

A provisional patent application can be filed up to one year following the date of first sale, offer for sale, public use, or publication of the invention.

These pre-filing disclosures, although protected in the U.S., may preclude patenting in foreign countries.

Unlike a non-provisional patent, the provisional patent is filed without any formal patent claims, oath, declaration, information disclosure, or prior art statement. The written description of the invention and any drawings necessary to more fully understand the invention must be provided for in an application for a provisional patent.

If either of these two items is missing or incomplete, your application will be rejected and no filing date will be given for your provisional application.

10.4.2 Advantages and Disadvantages

There are several advantages and disadvantages to provisional patents—the reader is well advised to seek competent patent counsel. A well-written provisional patent application should satisfy all formal and substantive legal requirements of the patent law. A provisional patent application should always be as complete as possible as compared to a non-provisional patent application; however, you will not be required to file any claims. You can conveniently use that grace year to collect additional data that may form a basis for eventual claims.

The entrepreneur/inventor must carefully balance the plusses and minuses of provisional patents, as shown in Table 10.4.

Table 10.4   Advantages/Disadvantages of provisional patents

Advantages	Disadvantages
Relatively simple and inexpensive. The specifications will not be examined. More comprehensive disclosures can be followed at later time.	A provisional application automatically becomes abandoned when its pendency expires 12 months after the provisional application filing date by operation of law.
Can claim priority to multiple provisional applications, as measured from the earliest filed application.	Provisional applications will not mature into a granted patent without further submissions by the inventor.
Once docketed, the term "Patent Pending" can be used for business purposes.	If not followed by a utility filing within one year, the inventor must cease using Patent Pending to avoid charges of false marking.
The effective patent term is 21 years from filing date of the provisional application.	Examination is delayed up to one year.

10.4.3 Patents Can Have Siblings/Children

Additional applications by the same inventor can be divided into three parts:

- Divisional
- Continuation
- Continuation-in-part

A **divisional** application is a type of patent application that contains matter from a previously filed application (the so-called parent application). While a divisional application is filed later than the parent application, it may retain its parent's filing date, and will generally claim the same priority.[5] A **continuation** application is a second (or subsequent) application directed to the same invention claimed in a prior co-pending application of the same inventor. A **continuation-in-part** is an application filed during the pendency of an earlier application of the same inventor, the latter application containing at least a substantial part of the disclosure of the earlier application together with an additional matter not so disclosed.

10.4.4 International Patent Organizations

The entrepreneur/inventor has three options when applying for a patent:

1. United States Patent Office (UPTO)
2. Patent Cooperation Treaty (PCT)—The PCT is an international agreement for filing patent applications having effect in up to 117 countries.[6] Although the PCT system does not provide for the grant of an international patent, the system:
   - simplifies the process of filing patent applications
   - delays the expenses associated with applying for patent protection in other countries and allows the inventor more time to assess the commercial viability of the invention
3. European Patent Office (EPO)—The EPO grants European patents for the Contracting States to the European Patent Convention. The EPO provides a single patent grant procedure, but not a single patent from the point of view of enforcement. EPO includes:
   - Major European countries
   - File in English, French, or German

10.5 Basis for Venture Investment

One of the major goals faced by seed and startup enterprises is access to initial or subsequent rounds of financing through either equity or debt. Investors are crucially interested where an innovation fits in the marketplace with reference to existing or potential competitors. Venture investors want to know if the innovation promises a sustained advantage; and a patentable innovation or invention offers compelling evidence to warrant a risky investment in the business, leading to eventual commercialization of the product.[7] This is shown in Figure 10.4.

Venture investors need to maximize returns and minimize risks. As shown in Figure 10.4, venture risks can be classified into four categories: (1) market, (2) financial, (3) management, and (4) technological.[8]

10.6 The America Invents Act

The **Leahy–Smith America Invents Act (AIA)** is a U.S. federal statute that was passed by Congress and was signed into law by President Obama on September 16, 2011. Named for its lead sponsors, Sen. Patrick Leahy (D-VT) and Rep. Lamar Smith (R-TX), the Act switches the U.S. patent system from a "first to invent" to a **"first inventor to file"** system, eliminates interference

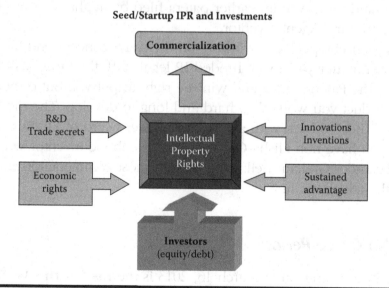

Figure 10.4   Seed/startup IPR—Investors are crucially interested if your innovation is patentable.

proceedings, and develops post-grant opposition. Its central provisions went into effect on March 16, 2013.[9]

The most sweeping changes to the U.S. patent system in over 50 years took effect on March 16, 2013, with the full enactment of the provisions of the *America Invents* patent reform act. Among other key changes in the current law, the act states that the first person to file a U.S. patent application will now be awarded the patent right, rather than the first to invent. Moreover, there will no longer be a grace period to file a U.S. application after any public disclosure or sale of an inventive product. It is important for all innovators, inventors, entrepreneurs, and business owners to be aware of, and to understand, the potential impact that *America Invents* will have on your intellectual property rights and business planning.

10.6.1 First to File

The biggest change is that the inventor who is **first to file** a patent application on an invention in the U.S. Patent Office now receives the patent right. Before AIA, an inventor could carefully document his or her discovery, conduct important technical and market research and, within reason, casually proceed with a patent filing when the product was ready and, more importantly, funds were available from investors and others to proceed with the costly act of patenting a product. As long as the inventor could prove he or she invented first, even an earlier patent filed by mighty GM or Microsoft would lose to our diligent inventor.

Now, if your competitor is working on the same concept and has a faster R&D team or, better yet, a well-funded IP legal staff, then they will likely beat you to the Patent Office and win the right to put you out of business in the very product you worked so hard and long to develop. The old "race to the Patent Office" is back with a vengeance. Ever heard of Elisha Gray? Most probably, you have not. Elisha Grey was the telephone inventor who lost the race to Alexander Graham Bell, even though most commentators agree he invented the telephone first.

10.6.2 No Grace Period

Another critical change after March 16, 2013 is the **loss of the 1-year grace period**. This allowed the inventor's product to be sold, used, or displayed in public for up to 1 year before the inventor actually needed to file a patent on it. You must now file a provisional patent application before publishing your results, should you wish to protect your patent rights.

Now all that sympathetic forgiveness in the U.S. patent system will be gone. If you publicly show/use a product or offer to sell it after March 16, 2013, then you will not be able to claim a patent on it unless your patent application was filed beforehand. There is a small exception for publications by the inventor, but it is best to consult your patent attorney before trying to use it.

Previously an inventor or startup could still obtain a patent after the product was already shown or on sale for a few months. This is now impossible. William Loginov, Esq.,[10] a patent attorney, developed a quick list of key points to help you survive this savage jungle of U.S. patent reform:

- Did you perform a patent search of your concept? This can streamline the filing process and avoid surprises later, especially if the patent is challenged by a competitor in "post-grant" proceedings.
- Did you consult a patent attorney early in the process? A patent can be filed as long as the basic concepts are workable. Waiting for actual "reduction to practice" is no longer a good idea in the race to file.
- Did you consider a provisional filing? Make sure it is well drafted as there is no grace period to fall back on if the provisional is ruled defective.
- Did you make sure to avoid any offers for sale, public displays, publications, or public use of the concept before consulting your patent attorney or filing a patent application? Otherwise, you may be kissing your rights goodbye.
- Did you get a non-disclosure agreement (NDA) for any needed discussions with third parties like vendors, suppliers, beta-testers, manufacturers, etc? An NDA makes these activities nonpublic, avoiding AIA. Remember, these third parties cannot do anything public with the concept; for example, any beta testing must be *strictly* internal and not for the purpose of serving paying customers.
- Are all your inventors and key employees under an NDA or similar agreement? Anyone with access to the concept should be under secrecy.
- As the project progresses, are you contacting the patent attorney at key development stages to update the patent coverage if needed? Remember this is now a *race*!

10.7 The Special Case Involving Biotechnology Patents

"Life sciences and biotechnology are widely regarded as one of the most promising frontier technologies for the coming decades."[11]

Research and development in the life sciences is extremely costly and time-consuming. The pharmaceutical industry provides a good example of the timescales (8 to 10 years) and funding to bring a drug or biological product to market. Most biotechnology startups do not have the financial resources available to them to survive for that period of time.

The business model of biotech firms often relies heavily on intellectual property rights, in particular patents, as they are often the most crucial asset they own in a sector that is extremely research-intensive and with low imitation costs. Investors in biotech companies are generally well aware of the centrality of patents and the survival of such companies may very well depend on their ability to convince investors that they have a solid IP strategy and that risks are reduced to a minimum.[12]

Why are patents so important for companies in the biotechnology sectors? This may be difficult to understand without understanding how the industry operates. According to Estevan Burrone, a consultant for Europe's Small and Medium Enterprises (SME) Division,[13] there are five major reasons:

1. Biotechnology is probably one of the most research-intensive industries. Compared with other major industries that also rely on R&D, such as the chemical industry, where the ratio of R&D expenditure to total revenues is approximately 5%, or the pharmaceutical industry, for which the equivalent figure is generally no more than 13%, biotechnology companies generally invest between 40% and 50% of their revenues in R&D. As in any research-based industry, the protection of research results becomes a major issue.
2. There are generally exorbitant costs for the development of new products and processes, but relatively low costs of imitation. The costs of performing biotechnology research are to be considered in the context of the high risks involved in any research project. It is hard to predict at the outset whether years of research will lead to breakthrough innovations with a great market potential or may simply leave a company empty-handed with results that are unlikely to bring revenues. Given the high costs involved in R&D, the relative ease of imitation is an issue

that is of great concern. According to the founders of Nordic Biotech, "the present reality in drug development (...) is that almost any technology or compound can rapidly be reverse engineered.[14]" Adequate IP protection becomes a means to ensure that biotechnology companies can appropriate their R&D results and reduce the likelihood of imitation by competitors.

3. Contrary to traditional industries, where there is a clear distinction between the basic research performed in universities and public sector R&D institutions on the one hand, and the applied R&D undertaken by private enterprises on the other, in biotechnology, basic and applied research are often profoundly inter-linked. Research undertaken in academic research institutions is often the basis for the establishment of biotechnology spin-offs. Similarly, biotechnology companies are often involved in (and are actively patenting) what some consider to be basic research.

4. The biotechnology industry, in most countries, consists mainly of recently established SMEs, an important number of which have yet to take a product to market. In many cases, biotechnology SMEs are established based on one or more patents developed within, or in partnership with, public research organizations or universities.

5. Finally, a point that derives from some of the issues discussed previously is that for some biotech companies intellectual property rights are actually the *final* product. It is not uncommon, in fact, to find biotechnology companies that develop innovative inventions, patent them, and then license them to larger companies that have the resources to take the product to market. Such companies may actually never sell a product themselves in the traditional sense, but base their revenues on their ability to develop, protect, and out-license their innovations.

In addition, biotechnology patents are a breed unto themselves. For example, you cannot obtain general patent protection for DNA sequences of a novel gene in a number of species if you have only sequenced a single vertebrate or invertebrate example. Description of the species usually does not allow protection for the genus in patents in the biosciences. In addition, DNA sequences for which no function has been demonstrated are generally not considered patentable. Many laws and regulations must be met, as shown in the following.[15]

10.7.1 Genetic Engineering Patents

Isolated DNA sequences, proteins to which functions have been attributed, and other metabolites are usually viewed in patent terms as chemical compounds, much like a new organic drug molecule. The unique sequence of the nucleotides or amino acids that you have uncovered constitutes a novel biological molecule (much like a novel chemical molecule) and may thus be patentable. In addition, vectors containing your nucleotide sequence and cells containing the vector/DNA may also be patented, provided they are considered new.

10.7.2 Microbiological Sciences Patents

Genetically modified organisms used in such processes may be eligible for patent protection. In addition, new microbes that you have isolated, purified, and cultured are generally considered patentable, provided they can fulfill the usefulness patent requirements.

10.7.3 Plant and Animal Sciences Patents

According to the patent laws of several countries, you cannot obtain biotechnological patent protection for plant or animal varieties, or essentially biological processes for the production of plants or animals. The U.S. is the exception to this and issues so-called Plant Patents. Similarly, biotech patent claims to animals obtained by traditional breeding methods are not allowable at most patent offices, but a genetically modified animal is considered patentable in the U.K., Europe, and the U.S. There is currently no equivalent in the animal sciences field to the protection offered by Plant Breeders' Rights.[16]

10.7.4 Pharmaceutical and Chemical Sciences Patents

Novel purified chemical or pharmaceutical compounds are patentable, as well as their pharmaceutically acceptable isomers and salts. Crude extracts in which a compound is enriched may also be patentable, depending on the level of enrichment relative to the natural, unfractionated state. Importantly, novel pharmaceutical carriers may also be patented. Patent protection may also be obtained for pharmaceutical compositions containing your novel pharmaceutical compound.

10.7.5 *Medical Sciences Patents*

Due to the medical patent restriction on methods of treatment, diagnosis, or surgery mentioned previously, surgical techniques are specifically excluded from patent protection in these regions. However, instruments for use in surgery, diagnosis, or therapy may be patented. In addition, diagnosis based upon a sample obtained from the body is allowable and should, accordingly, be limited to *in vitro* applications in a patent application.

10.7.6 *Microorganisms and Sufficiency of Description*

In the complex field of biotechnology, it is not always possible to fully describe a microorganism in terms of physical, chemical, and genetic characteristics in a patent specification. The Budapest Treaty[17] provides a solution to this patent problem—patent applicants may deposit a sample of the organism (as claimed in the patent specification) at a recognized patent depository and in doing so may overcome patent examiners' arguments as regards insufficiency of description of the microorganism in the biotech patent application. The deposit number of the sample must be reflected in the patent specification and the deposit must have been made before or at the time of filing the patent application.

10.8 Practical Advice to Entrepreneurs/Inventors

"I'm not going to buy my kids an encyclopedia. Let them walk to school like I did." —Yogi Berra

Patents do not merely protect inventions against imitators; they can also be used to block a competitor's technical progress. Thus, they are an integral part in your competitive advantage armamentarium.

These exclusive rights allow owners of intellectual property to benefit from the property they have created, providing a financial incentive for the creation of an investment in intellectual property, and, in the case of patents, attract investment capital.

The founder/entrepreneur needs to be aware of the time required to "prosecute" a patent, as shown in Figure 10.5.

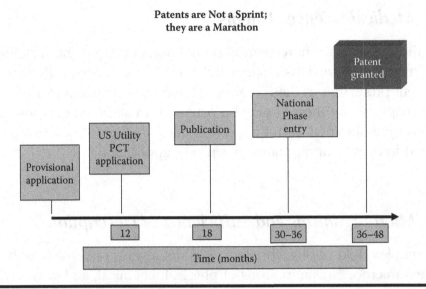

Figure 10.5 Patents are not a sprint—Being awarded a patent usually takes several years.

10.8.1 Who Is the "Inventor" in Your Company?

"Are you a starter or a finisher?"

Are you THE inventor? How about the other members of your team that helped you? Aren't they also considered inventors? Did not other members help you in "reduction to practice"? A "reduction to practice" is the embodiment of the concept of an invention by producing a physical incarnation of the invention. We offer four examples of reduction to practice: (1) for an item of manufacture, by providing a prototype or model of the invention; (2) for a method, by performing the procedure and experimentally confirming the results; (3) for a composition of matter, by synthesizing or otherwise producing the invention; or (4) by the very act of filing a patent application (also known as *constructive* reduction to practice).[18]

10.8.2 Conception

Conception is the cornerstone to determine inventorship.
Conception is the "formation in the mind of the inventor, of a definite and permanent idea of the complete and operative invention, as it is hereafter to be applied in practice."[19] Unless a person *participates* in the conception of

the invention, he or she *does not qualify* as an inventor. A complete conception is composed of two parts:

1. recognition of the ultimate desired end
2. development of the means to accomplish the desired result

This is important because U.S. patent law has the unique requirement that inventors must apply for a patent, and the inventors must be correctly named or the issued patent may be invalid. At the risk of getting too technical, 35 U.S.C. § 111 ("An application for patent shall be made, or authorized to be made, by the inventor…."); 35 U.S.C. § 115 ("The applicant shall make oath that he believes himself to be the original and first inventor…"), and 35 U.S.C. § 116 ("When an invention is made by two or more persons jointly, they shall apply for patent jointly…").

Note that once the inventor has been properly identified, the "inventor may use the services, ideas, and aid of others in the process of perfecting his invention without losing his right to a patent."[20]

10.8.3 *Should You Out-License Your Patent?*

License agreements contribute substantially to the income stream of many startup and medium-sized companies. Entering into an out-licensing agreement with a larger company may be entered by the licensor (innovator) with a licensee (strategic partner) for a variety of reasons including:

■ Lack of adequate production or distribution capabilities
■ Lack of financial and technical resources to further develop the invention
■ Manufacturing facilities may be prohibitively expensive
■ Lack of marketing muscle
■ Licensee needs the technology immediately
■ Cross-licensing is an important technique in settling patent infringement disputes

Please note that the term "license" is often, but incorrectly, used as the term for a fee paid by the licensee as the invention is sold in the marketplace. If this fee is tied to sales of the product, then the correct terminology for such a fee is a "royalty." Many licensing agreements are written containing both a fee and a royalty.

10.9 Fundraising Importance of Patents

Patents enable entrepreneurs to acquire financial capital under the most favorable terms in earlier founding rounds.[21] Firms having larger patent portfolios enjoy a greater likelihood of sourcing initial capital, and of achieving liquidity through an initial public offering. Given the lengthy government certification process, it provides a reliable "due diligence" signal for investors by which the quality of their investment can be quantified.[22]

Given the fundraising importance of patent protection, the entrepreneur/investor must guard against making innocent mistakes that may prevent his or her ability to obtain a patent. An inventor has one year from the date of "first public disclosure" of the invention to file a patent application in the U.S. Patent Office. "First public disclosures" are many and, surprisingly, they include:

■ Disclosures at presentations and poster sessions
■ Prior public uses or demonstrations
■ Prior conversations with potential partners
■ Prior publications or interviews
■ Prior sales of prototypes at beta sites

Also, keep in mind that if the U.S. application is filed after the public disclosure, the inventor is not able to file or patent protection in many countries. There is only a six-month "grace" period in Japan, Korea, and Russia. There is a one-year grace period in Canada, Australia, and Mexico.

Also, understand that the loss of foreign patent rights greatly diminishes the financial value of intellectual property in the eyes of investors. For that reason, and for your maximum protection, consider filing a patent application before embarking on any public disclosure.

■ Is the patent in a subject area that is earning significant profits?
■ Are there currently patent litigation cases in process in the subject area?
■ Does the invention allow for reduced costs or increased performance?
■ Are there any competitors that could directly benefit from your invention?
■ Are there blocking or dominating patents in this area?
■ Do you have freedom to operate?
■ Do you have an inventorship/ownership policy in place?

In summary, a patent can help your company be more "investable." Fundamentally, investors will analyze the risks and potential rewards of a single investment. Owning one or more patents can reduce the risk of the company by strengthening the competitive advantage and providing an additional marketable asset.

10.10 Trade Secrets in the StartUp Environment

Trade secrets are confidential and undisclosed business information that provides the owner with a competitive advantage. Virtually all types of information can be protected as a trade secret as long as reasonable and accepted measures are taken within the confines of the law. For example, the Uniform Trade Secrets Act, codified as Iowa Code Chapter 550, defines a trade secret as "information of nearly any kind that derives economic value from not being generally known or readily ascertainable by proper means, and is the subject of reasonable efforts to maintain its secrecy."

Examples of trade secrets include:

■ Software, including source code
■ Chemical formulations
■ Customer lists
■ Business plans
■ Pricing models
■ Marketing and sales strategies
■ Manufacturing costs

A startup needs to strongly consider whether to opt for patent protection or trade secret for certain innovations. Patent protection lasts for 20 years; a trade secret can last indefinitely unless (1) the information is no longer confidential, (2) the information has been discovered by legitimate means (such as reverse engineering), or (3) you do not maintain its confidentiality.

Thus, for certain innovations, the entrepreneur must decide *a priori* whether economically it will be more advantageous to seek a 20-year patent monopoly or to maintain the information as a trade secret potentially indefinitely.

References

1. Raysman, R., Pisacreta, E.A., & Adler, K.A. *Intellectual Property Licensing: Forms and Analysis.* Law Journal Press, 1998–2008.
2. *WIPO Intellectual Property Handbook.* WIPO Publication No. 489(E) 2004. http://www.wipo.int/export/sites/www/about-ip/en/iprm/pdf/ch2.pdf.
3. Maynard, J.T., & Peters, H.M. *Understanding Chemical Patents.* Washington, D.C.: American Chemical Society, 1991.
4. Modified after Maynard, J.T., & Peters, H.M. *Understanding Chemical Patents.* Washington, D.C.: American Chemical Society, 1991, p. 86.
5. Paris Convention for the Protection of Industrial Property. WIPO. http://www.wipo.int/treaties/en/ip/paris/trtdocs_wo020.html#P83_6610
6. http://www.epo.org/
7. Cardullo, M.W. Intellectual Property – The Basis for Venture Capital Investments. http://www.wipo.int/sme/en/documents/venture_capital_investments.pdf
8. Henos, M. Venture capital and intellectual property. *The Atlanta Journal /The Atlanta Constitution*, 1993.
9. http://en.wikipedia.org/wiki/Leahy-Smith_America_Invents_Act
10. Loginov, W., Esq. http://www.loginovlaw.com
11. European Commission. Life Sciences and Biotechnology – A Strategy for Europe, 2002.
12. Burrone, E. Patents at the Core: the Biotech Business.
13. http://www.wipo.int/sme/en/documents/patents_biotech.htm
14. Medicon Valley Patent Guide, 2002. Medicon Valley is a Danish/Swedish organization. http://www.mva.org/media(3,1033)/Medicon Valley Patent Guide.pdf and www.wipo.int/sme
15. Hoffelner, C. Patents—Biotech/Biotechnology. http://www.svw.co.za/patents-biotech.html
16. http://en.wikipedia.org/wiki/Plant_breeders'_rights
17. WIPO Biotechnology. http://www.wipo.int/patent-law/en/developments/biotechnology.html and www.wipo.int/patents/en/topics/biotechnology.html
18. http://www.patentlens.net/daisy/patentlens/2421.html
19. *Hybritech Inc. v. Monoclonal Antibodies, Inc.,* 802 F.2d 1367, 1376 (Fed. Cir. 1986).
20. *Shatterproof Glass Corp. v. Libbey-Owens Ford Co.,* 758 F.2d 613, 624 (Fed Cir. 1985).
21. Hsu, D., & Ziedonis, R.H. Patents as Quality Signals for Entrepreneurial Ventures. Copenhagen, DRUID Summer Conference, Denmark, June 2007. http://www2.druid.dk/conferences/viewpaper.php?id=1717&cf=9
22. Spence, M. Job market signaling. *Quarterly Journal of Economics*, 87, 355–374, 1973.

Chapter 11

Meet the JOBS Act

11.1 Introduction

In an effort to jump-start the entrepreneurial economy, the JOBS (Jumpstart Our Business Startups) Act created a new provision in the Securities Act of 1933 Section 4(6) that allows Emerging Growth Companies (EGCs) to raise up to $1 million in any 12-month period by selling securities through authorized intermediaries, subject to certain limitations on the matter of the offering and by limiting the amount any person is permitted to invest.

EGCs are a new category of issuer. EGCs are those with (1) less than $1 billion total annual gross revenues in their most recent fiscal year, and (2) have not had a registered public offering before December 8, 2012.[1]

The JOBS Act (signed into law April 6, 2012) facilitates financing across the spectrum from seed capital to public offerings. Following are some of the most important aspects and implications:

- Permitting "crowdfunding"
- Easing restrictions on fundraising from accredited investors
- Easing mandatory reporting triggers under the SEC Act
- Increasing the amount of money companies may raise in "mini-IPOs"
- Reducing many burdens on EGCs going public
- Providing more capital to entrepreneurs and EGCs, creating jobs and providing opportunities for non-accredited investors to invest in both community-based businesses and entrepreneurial companies

- For the last several years, the number of VC financings in the U.S. has continued to drop—approximately 3500 VC-led deals; VCs are raising less capital and continue to finance only larger opportunities with significant IRR potential and with exits of greater than $50 million
- Although Angel statistics are difficult to obtain, they funded nearly as much as VCs
- Fewer than 10% of all accredited investors in the U.S. invest in private financings; except as friends or family, non-accredited investors have no exposure to private financings
- There are 25,000,000 EGCs in the U.S.; many are looking for funding and banks are not lending; identifying investors is extremely difficult given securities laws

11.2 The JOBS Act at a Glance

The JOBS Act seeks to accomplish this goal by, among other measures, relaxing certain provisions of the Sarbanes-Oxley **and** Dodd-Frank Acts insofar as those provisions apply to a class of newly public companies dubbed EGCs. A primary goal of the legislation is to facilitate the ability of growing companies to raise capital, as follows:

- Removes the prohibition on general solicitation in connection with transactions dealing with Rule 508 or Rule 144A, provided that sales are limited to qualifying investors
- Allows the thresholds that trigger registration of a security under Section 12(g), including a different threshold for banks and bank holding companies
- Provides, to a new category of EGCs, relief from requirements and other restrictions applicable to IPOs and on a transitional basis for up to 5 years, relief from certain reporting requirements
- Adds a "crowfunding" exemption
- Authorizes the SC to increase the amount permitted to be raised in a Regulation A offering to $50 million in any 12-month period
- Modifications to Rule 506 will provide substantial freedom for issuers to promote their offerings to a wider group of investors
- Anyone who can convince the investing public that they have a good business idea can become an entrepreneur

- Modeled in part on campaign donations because politicians have been collecting small donations from the general public for decades
- Another route for business funding because VCs reject 98% of business plans

11.2.1 Title III of the U.S. JOBS Act

- The Act limits both the aggregate value of securities that an issuer may offer through a crowdfunding intermediary and the amount that an individual can invest.
- An issuer may sell up to an aggregate of $1,000,000 of its securities during any 12-month period.
- Investors with an annual income or net worth of up to $40,000 will only be permitted to invest $2,000 and above $40,000 and less than $100,000, investors shall be entitled to invest 5% of their annual income or net worth in any 12-month period.
- Investors with an annual income or net worth greater than $100,000 will be permitted to invest 10% of their annual income or net worth.
- Investors are limited to investing $100,000 in crowdfunding issues in a 12-month period.
- Investors who purchase securities in a crowdfunding transaction are restricted from transferring those securities for a period of one year. This restriction is subject to certain exceptions, including transfers: (i) to the issuer; (ii) to an accredited investor; (iii) pursuant to an offering registered with the SEC; or (iv) to the investor's family members.

11.2.2 Equal Access and Disclosure

Equal access to and disclosure of material information is a core principle of federal and state securities regulations. It is essential for investors to have the necessary information to appreciate the potential risks and rewards of an investment. The JOBS Act requires issuers to provide investors with a description of the following:

- Company: the issuer and its members, including the name, legal status, physical address, the names of the directors and officers holding more than 20% of the shares of the issuer.
- Offering: the anticipated business plan of the issuer, the target offering amount, the deadline to reach the target offering amount and the price to the public of the securities.

- Structure: the ownership and capital structure of the issuer, including terms of the securities of the issuer being offered.
- Valuation: how the securities being offered are being valued, and examples of methods for how such securities may be valued by the issuer in the future, including during subsequent corporate actions.
- Risks: the risks to purchasers of the securities relating to minority ownership in the issuer, the risks associated with corporate actions, including additional issuances of shares, a sale of the issuer or of assets of the issuer, or transactions with related parties.

The intermediary crowdfunding portals are also required to make available to the SEC and to potential investors any information provided by the issuer no later than 21 days prior to the first day on which securities are sold to any investor.

11.3 Crowdfunding

Crowdfunding refers to the funding of an EGC by selling small amounts of equity to many investors. This form of crowdfunding has recently received attention from policymakers in the U.S. with direct mention in the JOBS Act, legislation that allows for a wider pool of small investors with fewer restrictions.[2]

With the passing of the Act, the word of the day seems to be crowdfunding. While this concept has arguably been around a long time, it is still formally recognized as a new industry to many consumers, particularly those outside the U.S. **Crowdfunding** is, by definition, "the practice of funding a project or venture by raising many small amounts of money from a large number of people, typically via the Internet."

Crowdfunding has its origins in the concept of crowdsourcing, which is the broader concept of an individual reaching a goal by receiving and leveraging small contributions from many parties. Crowdfunding is the application of this concept to the collection of funds through small contributions from many parties in order to finance a particular project or venture.[3]

Theoretically, crowdfunding allows EGCs to sell securities to anyone, without being compelled to produce the onerous amounts of information currently required by existing federal law. A number of U.S. organizations have been founded to provide education and advocacy related to equity-based crowdfunding as enabled by the JOBS Act. They include:

- National Crowdfunding Association
- Crowdfunding Professional Association
- CrowdFund Intermediary Regulatory Advocates

Crowdfunding is not available to non-U.S. companies, public companies, or investment companies, including companies exempt by Section 3(b) or 3(c) of the Investment Company Act of 1940. In addition, securities sold in a crowdfunding deal may not be transferred for one year from the date of purchase, except in limited circumstances.

11.4 Issuer Requirements

EGCs seeking to raise capital under Section 4(6)[4] are required to provide certain information to potential investors, such as

- the company
- its business
- officers and directors
- major stockholders (greater than 20%)
- terms of the offering securities being offered for sale

Importantly, the JOBS Act requires that EGCs must provide more detailed financial disclosures for larger offerings. Thus, if the aggregate amount of the offering is $100,000 or less, the issuer must only provide tax returns for the company's most recently completed fiscal year, and financial statements certified by the company's Chief Executive Officer.

In contrast, if the aggregate amount is $100,000 to $500,000, the issuer must provide financial statements reviewed by an independent public accountant. If the aggregate amount being offered exceeds $500,000, the issuer must provide audited financial statements.

11.5 Intermediary Requirements

The JOBS Act requires that crowdfunded offerings be conducted through authorized third-party "intermediaries."[5] Intermediaries, crowdfunding brokers, and funding portals have significant duties under the JOBS Act to provide information to investors, reduce the risk of fraud, and, where required under the Act, ensure that investors and issuers satisfy the requirements outlined in Title III of the JOBS Act.

The JOBS Act requires these intermediaries to, among other things:

■ provide disclosures that the SEC determines appropriate by rule, including regarding the risks of the transaction and investor education materials
■ ensure that each investor: (1) reviews investor education materials; (2) positively affirms that the investor understands that the investor is risking the loss of the entire investment, and that the investor could bear such a loss; and (3) answers questions that demonstrate that the investor understands the level of risk generally applicable to investments in startups, emerging businesses, and small issuers and the risk of illiquidity
■ take steps to protect the privacy of information collected from investors
■ take such measures to reduce the risk of fraud with respect to such transactions, as established by the SEC, by rule, including obtaining a background and securities enforcement regulatory history check on each officer, director, and person holding more than 20% of the outstanding equity of every issuer whose securities are offered by such person
■ make available to investors and the SEC, at least 21 days before any sale, any disclosures provided by the issuer
■ ensure that all offering proceeds are only provided to the issuer when the aggregate capital raised from all investors is equal to or greater than a target offering amount, and allow all investors to cancel their commitments to invest
■ make efforts to ensure that no investor in a 12-month period has purchased crowdfunded securities that, in the aggregate, from all issuers, exceed the investment limits set forth in section Title III of the JOBS Act; plus any other requirements that the SEC determines are appropriate

11.6 Funding Portals

Title III of the JOBS Act adds new Section 3(h) to the Exchange Act, which requires the SEC to exempt, conditionally or unconditionally, an intermediary operating a funding portal from the requirement to register with the SEC as a broker.

The intermediary, though, would need to register with the SEC as a funding portal and would be subject to the SEC's examination, enforcement, and rulemaking authority. The funding portal also must become a member of a national securities association that is registered under Section 15A of the Exchange Act.

A funding portal is defined as a crowdfunding intermediary that does not: (1) offer investment advice or recommendations; (2) solicit purchases, sales, or offers to buy securities offered or displayed on its website or portal; (3) compensate employees, agents, or other persons for such solicitation or based on the sale of securities displayed or referenced on its website or portal; (4) hold, manage, possess, or otherwise handle investor funds or securities; or (5) engage in such other activities as the SEC, by rule, determines appropriate.

The JOBS Act directs the SEC to adopt rules to implement Title III within 270 days of enactment of the Act. The President signed the JOBS Act into law on April 5, 2012.

11.6.1 *Restrictions on Funding Portals*

The JOBS Act imposes several restrictions on the activities of a registered funding portal. A funding portal is *not* permitted to:

■ provide investment advice or make recommendations
■ solicit purchases, sales, or offers to buy the securities offered or displayed on its website or portal
■ compensate employees, agents, or other persons for such solicitation or based on the sale of securities displayed or referenced on its website or portal
■ hold, manage, possess, or otherwise handle investor funds or securities
■ engage in any other activities the SEC determines to prohibit in its crowdfunding rulemaking

In addition, each funding portal and each crowdfunding broker is prohibited from:

■ compensating promoters, finders, or lead generators for providing the intermediary with the personal identifying information of any potential investor
■ allowing its directors, officers, or partners (or any person occupying a similar status or performing a similar function) to have a financial interest in any issuer using the services of the intermediary

11.7 Mini-IPOs

The Act also makes it easier for a company to raise a more substantial amount of capital without becoming an SEC reporting company by amending Section 3(b) of the Securities Act to exempt from registration any class of equity, debt, or convertible debt securities sold in an offering where the aggregate offering amount in any 12-month period does not exceed $50 million. This assumes that the issuer files audited financial statements each year following the offering and complies with any other rules to be developed by the SEC.

This amendment effectively expands the existing exemption under Regulation A, which allows a private company to raise up to $5 million from the public using a more streamlined disclosure than a typical PO and without registering the offering with the SEC or becoming an SEC reporting company.

11.8 Emerging Growth Company IPOs

Last, the JOBS Act makes it easier for EGCs to raise capital in the public markets by exempting them from a number of current and proposed regulations applicable to companies contemplating an IPO, and once they are public. The idea is to stimulate the economy by offering fledgling companies a relatively easy entry into the IPO thruway.

An EGC is defined as any issuer that had total annual gross revenues of less than $1 billion in the last fiscal year. An EGC retains that status for up to five years after its IPO (it may lose that status earlier if revenues reach more than $1 billion, if it issues more than $1 billion in debt, or floats more than $700 million in stock). Companies that went public before December 8, 2011 do not qualify for EGC status.[6]

Under the JOBS Act, EGCs would have to report only two years of audited financial statements when they go public (as opposed to three years for income statements, and five years for select financial data under current law). Underwriters participating in an EGC's IPO are able to issue research reports on the EGC's stocks ahead of offerings, a practice that is currently prohibited. Additionally, EGCs will be exempt from certain disclosure requirements regarding executive compensation.

The bill received bipartisan support, passing with a 390 to 23 vote in the House of Representatives. Supporters believe the bill targets rules that are impediments to growth, and makes it easier for young companies to raise capital and conduct IPOs. In a March 7 Senate Hearing, Senator Chuck

Schumer specifically cited the "drastic decline" in U.S. IPOs since the 1990s as a driving force behind the bill.

Critics also worry about loosening restrictions on the ability for underwriters to publish research reports about an issuer it is representing. After the dot-com boom, where analysts touted tech stocks that turned out to be bad investments, regulators believed that those analysts were influenced by their investment banking colleagues who were seeking to win business from those issuers. Sarbanes-Oxley aimed to reduce the influence of bankers on equity research, and critics fear that a rollback could have severe consequences for investor protection.

11.8.1 Relief for EGCs

As we have seen, the JOBS Act created a new category of EGCs, and provided various forms of relief aimed at making it easier for these companies to undertake IPOs.[7] An EGC is defined as any issuer that had total annual gross revenues of less than $1 billion during its most recently completed fiscal year, other than a company that completed its IPO on or before December 8, 2011.

An EGC retains that status until the earliest of:

- the last day of the fiscal year during which it had total annual gross revenues of $1 billion or more
- the last day of the fiscal year following the fifth anniversary of the issuer's IPO
- the date on which the issuer has, during the previous three-year period, issued more than $1 billion in non-convertible debt
- the date on which the issuer is deemed to be a "large accelerated filer," as defined in Rule 12b-2 under the Exchange Act (*i.e.*, has been a reporting company for 12 months, has filed at least one annual report, and has a market value of equity securities held by non-affiliates of $700 million or more as of the most recently completed second fiscal quarter).
- the $1 billion annual gross revenue threshold is to be adjusted for inflation every five years

The JOBS Act also exempts EGCs from the following additional compensation-related disclosure provisions that were imposed on U.S. public companies pursuant to the Dodd-Frank Act:

- the advisory "say-on-pay" vote on executive compensation required under Section 14A(a) of the Exchange Act
- the Section 14A(b) requirements relating to shareholder advisory votes on golden parachute compensation
- the Section 14(i) requirements for disclosure relating to the relationship between executive compensation and financial performance of the issuer
- the requirement of Dodd-Frank Act Section 953(b)(1), which will require disclosure as to the relationship between CEO and median employee pay

11.8.2 Confidential Filing Process

Under the JOBS Act, the SEC is required to permit EGCs to submit draft IPO registration statements on a confidential basis. The ability to file on a confidential basis could benefit many issuers by allowing them to delay disclosure of competitive or otherwise sensitive information until the issuer is reasonably sure that its IPO will proceed as planned. The ability to file on a confidential basis can also stave off the embarrassment often associated with the withdrawal of a registration statement due to lack of investor interest.[8]

Under the new confidential filing process, the confidential draft registration statement submitted to the SEC must be "substantially complete" and must include a signed audit report and exhibits, but does not need to be signed by the issuer or include an auditor consent. The submission of a confidential draft registration statement is not deemed a filing for most purposes under the Securities Act unless and until the registration statement is filed on a non-confidential basis, although such filing may trigger certain filing requirements for FINRA purposes. Although an issuer is prohibited from publicly announcing the fact that a confidential submission has been made, it is allowed to inform Qualified Institutional Buyers (QIBs) and Institutional Accredited Investors (IAIs) of such submissions pursuant to the "testing-the-waters" communications discussed next. An issuer is also permitted to publicly announce its intention to engage in a public offering.

The registration statement does not need to be publicly filed until 21 days prior to the first roadshow, or 21 days before anticipated effectiveness of the registration statement, if no road show is to be held. Communications with QIBs and IAIs under the test-the-water provisions of the JOBS Act will not be deemed to constitute road shows. Upon the initial filing of a registration statement on a non-confidential basis, all previously filed confidential submissions must be filed as an exhibit to the publicly filed registration statement. Accordingly, even confidential submissions must be drafted with an eye toward eventual public disclosure (assuming the IPO is pursued).

11.8.3 Relaxed Restrictions on Investor Communications

The JOBS Act also eases restrictions on communications with investors before and after the registration statement becomes effective.

Generally, the solicitation of orders to buy securities before a registration statement becomes effective (sometimes referred to as "gun-jumping") is prohibited under the Securities Act. However, under the JOBS Act, EGCs and their authorized representatives will be permitted to communicate orally or in writing with QIBs and IAIs to determine interest, or "test-the-waters," in a potential offering, whether before or after the filing of a registration statement for the offering. Such communications may only be with these types of investors; communications with any investors that do not qualify as QIBs or IAIs remain subject to the traditional prohibitions on gun-jumping. This new "testing-the-waters" exception is intended to allow an issuer to gauge interest in an offering prior to undertaking the significant expense of initiating the IPO process. Testing-the-waters communications are not limited to the EGC's IPO, but are also permitted during subsequent offerings, if the issuer still meets the qualifications of an EGC at such time. Issuers should be aware, however, that these activities will only be preempted from state blue sky laws if the securities will be "covered" securities when issued, that is, listed on the NYSE, AMEX, or Nasdaq.

In addition to testing-the-water communications, the JOBS Act loosened the restriction on communications by research analysts. Under the JOBS Act:

■ Investment banks will be permitted to publish research on an issuer during the pendency of such issuer's public offering, even if the bank is also serving as an underwriter in the offering
■ The research analyst conflict of interest rules related to marketing of IPOs and "three-way" communication between research, investment banking, and management will not apply
■ There will be no post-pricing quiet period or booster shot restrictions on research reports or other communications

These changes are intended to increase the availability of analyst research coverage of EGCs that may not otherwise receive attention from the investment community. These provisions override certain restrictive provisions of FINRA and the NYSE, but other Securities Act, FINRA, and NYSE restrictions will continue to apply.

Please note that despite these changes instituted by the JOBS Act, disclosures during the IPO process will still be subject to existing regulations against fraud. Accordingly, issuers are advised to take care that any disclosures during this period are consistent with the information provided in the IPO submissions and filings. This potential for liability, especially in light of other continuing restrictions, may limit the overall impact of these provisions.

References

1. http://ww2.cfo.com/growth-companies/2013/10/the-jobs-act-crowdfunding-and-emerging-businesses/
2. http://www.forbes.com/sites/tanyaprive/2012/11/27/what-is-crowdfunding-and-how-does-it-benefit-the-economy/
3. http://en.wikipedia.org/wiki/Crowdfunding
4. http://en.wikipedia.org/wiki/Jumpstart_Our_Business_Startups_Act
5. http://www.sec.gov/divisions/marketreg/tmjobsact-crowdfundingintermediariesfaq.htm
6. Xia, E. The JOBS act: risks for investor protection. *Columbia Business Law Review.* http://cblr.columbia.edu/archives/12083
7. http://blogs.law.harvard.edu/corpgov/2012/03/29/congress-passes-the-jumpstart-our-business-startups-act/
8. http://www.troygould.com/index.cfm?fuseaction=content.contentDetail&id=9206

Chapter 12

Strategic Management

12.1 Introduction

From the start, your immediate task is to identify and establish your company's sustainable strategic competitive advantage.[1,2] Your **strategy** is your business approach to a set of competitive moves designed to generate a successful outcome for your enterprise. It is your *game plan* for:

- Strengthening your organization's competitive position
- Satisfying customers
- Achieving your performance targets

You have three generic strategies available to you: (1) focus, (2) differentiation, and (3) cost leadership, as shown in Figure 12.1.

A **focused** (or market niche) outcompetes rivals by offering innovative (customized attributes) products that meet requirements/tastes not previously available. A **differentiated** strategy presents the market with enabling (revolutionary) new products that appeal to a broad spectrum of buyers. Last, a **cost leadership** strategy seeks to give customers more "value for their money" by offering good-to-excellent attributes at a lower cost than competitors offer, in price-sensitive markets.

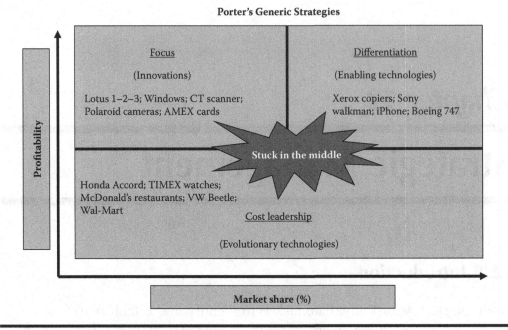

Figure 12.1   Porter's generic strategies—The three generic strategies to establish your sustainable advantage.

12.2 Strategic Corporate Planning

"The essence of strategy lies in creating tomorrow's competitive advantage faster than competitors mimic yours." —Hamel and Prahalad

As a founder, one of your most important strategies is to set the tone for your company (also known as culture). You need to develop a strategic plan to grow your company and align your organization as cross-functional teams. Your company will be a reflection of you.

Strategic corporate planning is the organization's process of defining its strategy, or direction, and making decisions on allocating its resources to pursue this strategy, in order to determine the future direction of the organization.[3] You will need to clearly communicate your strategy to your team. This will allow executives to make the hard decisions that come with a rapidly growing business, and provide you with immediate and timely feedback when crises occur.

Strategic planning consists of the following steps, summarized in Figure 12.2.

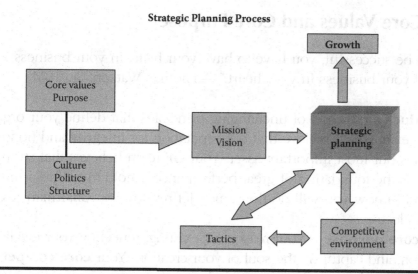

Figure 12.2 Strategic planning process—Strategic planning consists of three inputs.

Your decisions will determine which strategies are formulated and implemented to put your business model into action:

Goals (understood by all members)			
Founder	*Core values*	*Core principles*	*Core competencies*
Corporate	Mission, vision	Strategy	Tactics
Innovation R&D	Value creation Differentiation	Align with strategy	Product Launch
Finance	Fundraising Debt	Metrics of success	Cash flow Capitalization
Marketing Sales	Pricing Positioning	Market share targets	Product extensions Customer feedback
Personnel	Hire top talent	Hiring Firing	Empowerment Morale
Manufacturing	Outsourcing Geographical location	Production costs Inventory control	Value added Flexible production

12.3 Core Values and Core Purpose

"To be successful, you have to have your heart in your business, and your business in your heart." —Thomas Watson, Sr., IBM

Core values are the set of unchanging principles that define your organization. For example, "I believe that our reputation for integrity and honorable dealings is our most important asset. Your GE team believes that strong integrity is the foundation of great performance. I hold myself to a high standard, and I know you will do the same." Jeffrey Immelt, Chairman & CEO, General Electric.

The **core purpose** is your reason for existing, reflecting your idealistic motivation, and capturing the soul of your creation. Your **core competencies** are those things your company can do better than anyone else. For example:

■ Nike's core purpose is to experience the emotion of competition, winning, and crushing competitors.
■ 3M's core purpose is to solve unsolved problems innovatively.
■ Walt Disney's core purpose is to make people happy.
■ Apple's core competency is disruptive innovations.
■ Charles Schwab's core competency is lightning fast, cost-effective trades.
■ Aspen Dental's core competency is your mouth is our business.

Table 12.1 summarizes the two core principles you should inculcate in your organization, while is it still in the formative stages.

12.4 Mission, Vision

"You must sell yourself before you sell your company."

A **mission** is a compelling, daring, and achievable goal. Like any goal, a mission must be finite, measurable, and practical. Once achieved, a new mission must be set and articulated.

A **vision** statement is very broad and forward-looking and should be inspiring to employees as well as all other stakeholders. The vision statement clearly states what the company is all about and explains direction and purpose. As opposed to mission statements, vision statements unite the entire staff toward the pursuit of one common goal and are permanent.

Table 12.1   Core principles (Values purposes)

Core Values	Core Purposes
HOW?	WHEN?
Do we behave?	Do we measure our progress?
Do we achieve collective commitment?	Do we establish targets and timelines?
Do we guide behavior?	Do we establish priorities?

Collins and Lazier[4] argue that if you examine the historical evolution of great companies like L.L. Bean, Hewlett-Packard, Merck, Johnson & Johnson, Medtronic, Wal-Mart, Boeing, PepsiCo, Sony, McDonald's, etc. you will find that while the company was still relatively small, key leaders instilled a compelling vision into the organization.

For example, Johnson & Johnson's famous mission statement reads as follows:[5]

> We will delight our consumers, treating each person who contacts us as if they are our only consumer, providing them with a response which is evidence of our interest and that leaves them with the clear understanding that they are important to us.
> Our Johnson & Johnson Consumer Service Centre staff are committed to providing exceptional service to our consumers.
>
> - We will be contactable 24 hours a day, 7 days a week either personally or by voicemail.
> - We will be empowered to resolve any dissatisfaction that consumers have about our products and services.
> - We will send a response, where appropriate, to email and telephone contact within two working days and written contact within three working days of its receipt.
> - We will compose our response letters so they are easily understood.
> - We will provide consumers with replacement or alternative product or monetary refund in instances of product dissatisfaction.
> - In some cases, you will be asked to return the product (at our expense) so that we can conduct an appropriate review and take any necessary corrective action to the monitoring procedures in an endeavor to eliminate a repetition of the problem

in the future. Also, it is important that we better understand your requirements to assist us in delivering products that meet your needs.

■ You may be assured of our continued commitment to provide you and our other consumers with quality products and an exceptional level of service at all times.

12.5 Competitive Intelligence

"Never do anything you'd be afraid to read about on the front cover of the newspaper the following day." —Karen Rothwell, Outward Insights

Competitive intelligence (CI) is the intimate know-how of the competitive environment. In well-managed forms, CI is the prelude to strategic decision and action. The **competitive intelligence process** is the organizational means by which competitive information (the environment) is systematically collected, analyzed, processed, and disseminated as actionable intelligence to selected executives and managers "in need to know."[6]

Note that **information** is factual. It is numbers, statistics, bits of data, and interesting stories that seem important. On the other hand, **intelligence** is an *actionable* list of data that has been analyzed, filtered, and distilled. This is what we call intelligence know-how.

CI is not for everyone. In fact, most companies do not use CI. Despite the fact that most executives rely on the flow of information for decision-making, only a handful of companies have a fully functional, integrated CI process in place. Why is that? Perhaps the most important reason is attitudinal: the way executives think about CI.[7] Have you ever heard some of comments listed below? On the other hand, perhaps you have said them yourself!

■ CI is spying; it is unethical. I don't want any part of it.
■ It was not part of my school curriculum. It must not be important.
■ CI is a cost center. We do not have a budget for it.
■ How do I quantify CI's cost/benefit ratio?
■ Nothing happens in this industry that I don't know already.
■ If I don't know it, is it not worth knowing.
■ We tried it before, and it didn't work.
■ I am too busy to review all this garbage.

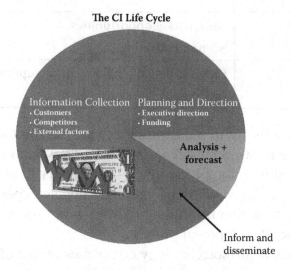

Figure 12.3 The CI life cycle—Competitive intelligence (CI) requires authorization and guidance from the highest echelons of management.

CI requires authorization by the highest echelons of management. CI is a top management function. It can provide vital analysis of competitor capabilities, plans, intentions, and limitations. It spotlights industry structure and trends. It may also reveal political, economic, and social forces affecting your company.

The CI life cycle is iterative, consisting of four major functions: (1) planning and direction, (2) information collection, (3) analysis and forecast, and (4) information and dissemination, as shown in Figure 12.3.

Note that CI is not merely a large collection of information; it must be analyzed and presented in actionable form (intelligence know-how). CI is most useful when presented in a forward-looking, predictive, and actionable format such as the following:

Competitive Intelligence (CI) Process	
CI is not a function; it is a process, or relegated to one area, division, or unit.	
Method	*Reason*
Strategic	Predict how competitors are likely to react
	Future intentions
Tactical	Predict price or new positioning by likely competitors
	Benchmarking

Offensive	New market opportunities
	Scenario planning
Defensive	Early warning of enabling or disruptive technologies
	Investment decisions

CI is particularly useful in industries with long development and approval cycles, such as the pharmaceutical and biotechnology industries. These industries are faced with long R&D times—sometimes 10 to 15 years for innovative drugs—coupled with uncertain and hugely expensive clinical trials. This allows continual tracking of a drug's progress by competitors through the public FDA approval process.

Thus, pharmaceutical/biotechnology companies utilize CI during the years of drug development to help determine if a new drug development should be continued or dropped. Likewise, it allows companies to monitor competitors' activities and decide whether to initiate their own drug development for a specific indication.

12.6 Marketing Intelligence

"You can see a lot by watching." —Yogi Berra

Marketing intelligence (MI) is the information relevant to a company's markets, gathered and analyzed specifically for the purpose of accurate and confident strategic decision-making in determining market opportunity, market penetration, true market size, market strategy, and market development metrics.[8] MI is the systematic gathering of data from the company's current external environment to further a company's sales goals.

The four cornerstones of MI are (1) competitor intelligence, (2) product intelligence, (3) market intelligence, and (4) environment intelligence,[9] as seen in Figure 12.4.

While the microenvironment refers to those elements closest to the company, customers, competitors, suppliers, etc., the microenvironment refers to those elements that can affect a company, but cannot be controlled by management. These include external factors such as the economy, culture, demographics, politics, technology, and weather.

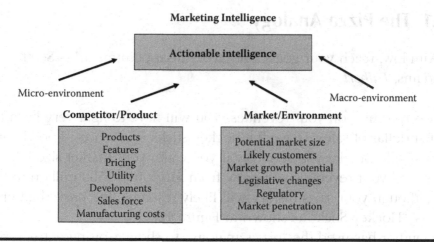

Figure 12.4   Marketing intelligence—The four cornerstones of marketing intelligence.

MI provides management with a dispassionate look at their target market. Data eats expectation for lunch. Founders generally have a distorted look at their selected market, confusing their total available (global) market with their target (niche) market, which is a much smaller percentage than they anticipated. (Note that actual market is equivalent to market share.) This is shown in Figure 12.5.

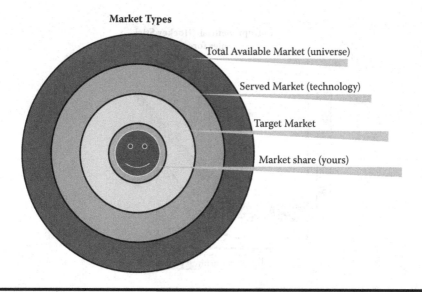

Figure 12.5   Market types—What is your real potential market share? No baloney, please.

12.6.1 The Pizza Analogy

"Aim low, reach your goals, and avoid disappointment." —Scott Adams, *Dilbert*

Don't overpromise on your revenues. You will have to fight very hard for your first dollar of sales. Leave the hockey sticks on the ice—knowledgeable investors will not take you seriously if you claim your market size is in the billions, and your revenues will grow from $10,000 to $500 million in three years, although your sales history will likely resemble the proverbial entrepreneurs' "Hockey Stick" as shown in Figure 12.6.

The author has used the "pizza analogy" to show aspiring entrepreneurs the intricacies of market/sales terminology. Suppose you have a complete pizza, as shown in Figure 12.7.

The entire 8-slice pizza is the **Total Available Market**. However, your technology can only address half of the market; that is, 4 slices (Figure 12.8).

The 4 pizza slices now represent your **Served Market**. If you could somehow eat 100% of the served market, you would walk on water. Realistically, you could only expect to eat 25% of the served market, or one slice. This is equivalent to your *potential* **Target Market**, as shown in Figure 12.9.

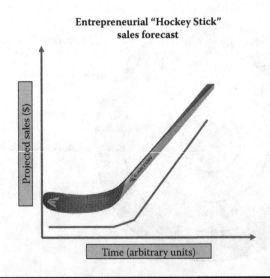

Figure 12.6 **Entrepreneurial hockey stick—Will your revenues truly resemble the stick part of a hockey stick?**

The Total Available Market
(8 slices)

Figure 12.7 The whole pizza pie—Your total available market.

Your Served Market

Figure 12.8 Technology reach—Your technology is only capable of addressing "4 slices" or Served Market.

Your Target Market (1 slice)

Figure 12.9 Your target market—Upon examination, your realistic market is the Target Market (1 slice).

12.7 Strategic Marketing Goals by MI

"We made too many wrong mistakes." —Yogi Berra

Most entrepreneurial firms are moving away from "mass" marketing to "target" marketing. Mass marketing is the production approach of identifying what most consumers want, and assumes actions of all consumers are equal, based on mass production (low-cost, effective, easy).

Entrepreneurial firms increasingly focus on target marketing, where the consumer needs set the product characteristics and production efforts change to meet needs of *targeted* groups of consumers. Strategic marketing goals lead to premium pricing and low production volumes because you offer what others cannot.

According to Seperich,[10] one way to study the market and marketing goals is to look at the chances or possibilities for market growth and projected market share, and clearly identify the best products a firm is offering.[11] This is shown in Figure 12.10.

As shown in Figure 12.10 according to the Boston Consulting Group (BCG) growth-share matrix, business could be classified as high or low according to their industry growth rate and relative market share. The analysis requires that both measures be calculated for each Strategic Business

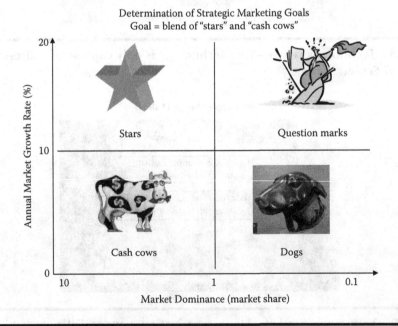

Figure 12.10   Determination of strategic—The famous BCG "growth-share matrix."

Unit (SBU). The dimension of business strength—relative market share—will measure comparative advantage indicated by market dominance. The key theory underlying this is existence of an experience curve and that market share is achieved due to overall cost leadership.

The BCG matrix has four cells, with the horizontal axis representing relative market share and the vertical axis denoting market growth rate. The mid-point of relative market share is set at 1.0. If all the SBUs are in the same industry, the average growth rate of the industry is used. However, if all the SBUs are located in different industries, then the mid-point is set at the growth rate for the economy.

Resources are allocated to the business units according to their situation on the grid. The four cells of this matrix have been called stars, cash cows, question marks, and dogs. Each of these cells represents a particular type of business, as described here:[12]

1. **Stars:** Stars represent business units having large market share in a fast growing industry. They may generate cash but because of fast growing market, stars require huge investments to maintain their lead. Net cash flow is usually modest. SBUs located in this cell are attractive as they are located in a robust industry and these business units are highly competitive in the industry. If successful, a star will become a cash cow when the industry matures.

2. **Cash Cows:** Cash cows represent business units having a large market share in a mature, slow growing industry. Cash cows require little investment and generate cash that can be utilized for investment in other business units. These SBUs are the corporation's key source of cash, and are specifically the core business. They are the base of an organization. These businesses usually follow stability strategies. When cash cows lose their appeal and move toward deterioration, then a retrenchment policy may be pursued.

3. **Question Marks:** Question marks represent business units having low relative market share and are located in a high-growth industry. They require huge amounts of cash to maintain or gain market share. They require attention to determine if the venture can be viable. Question marks are generally new goods and services that have a good commercial prospective. No specific strategy can be adopted. If the firm thinks it has dominant market share, then it can adopt expansion strategy, or else retrenchment strategy can be adopted. Most businesses start as question marks as the company tries to enter a high-growth market in

which there is already a market share. If ignored, then question marks may become dogs, while if huge investment is made, they have the potential of becoming stars.

4. **Dogs:** Dogs represent businesses having weak market shares in low-growth markets. They neither generate cash nor require huge amounts of cash. Due to low market share, these business units face cost disadvantages. Generally, retrenchment strategies are adopted because these firms can gain market share only at the expense of competitor or rival firms. These business firms have weak market share because of high costs, poor quality, ineffective marketing, etc. Unless a dog has some other strategic aim, it should be liquidated if there are fewer prospects for it to gain market share. The number of dogs should be avoided and minimized in an organization.

The BCG matrix produces a framework for allocating resources among different business units and makes it possible to compare many business units at a glance. However, the BCG matrix is not free from limitations, such as the following:

1. BCG matrix classifies businesses as low and high, but generally, businesses can be medium as well. Thus, the true nature of business may not be accurately reflected.
2. Market is not clearly defined in this model.
3. High market share does not always lead to high profits. There are high costs involved with high market share.
4. Growth rate and relative market share are not the only indicators of profitability. This model ignores and overlooks other important indicators of profitability.
5. At times, dogs may help other businesses in gaining competitive advantage. They can earn even more than cash cows sometimes.
6. This four-celled approach is considered too simplistic.

References

1. Porter, M.E. How competitive forces shape strategy. *Harvard Business Review*, March/April 1979.
2. Porter, M.E. *Competitive Advantage*. New York: Free Press, 1985.
3. http://en.wikipedia.org/wiki/Strategic_planning

4. Collins, J.C., & Lazier, W.C. *Beyond Entrepreneurship*. Englewood Cliffs, NJ: Prentice Hall, 1982.
5. http://www.jnjnz.co.nz/mission-statement
6. Best practices in managing a CI function, http://units.sla.org/division/dpht/meetings/annual2007/2007best-overview.ppt.
7. Kahaner, L. *Competitive Intelligence*. New York: A Touchstone Book, 1996.
8. http://en.wikipedia.org/wiki/Market_intelligence
9. Crowley, E. A market intelligence primer. http://www.pragmaticmarketing.com/resources/a-market-intelligence-primer.
10. Seperich, G.J., Woolverton, M.W., & Beierlein, J. *Introduction to Agribusiness Marketing*. Englewood Cliffs, NJ: Prentice Hall, 1994.
11. Boston Consulting Group growth-share matrix.
12. Easy matrix diagrams. http://www.managementstudyguide.com/bcg-matrix.htm.

Chapter 13

Valuation Techniques

13.1 Introduction

Every founder believes in his heart of hearts that his venture is truly worth a fortune, and it will be funded. From the founder's perspective, it should be obvious to every investor what an outstanding opportunity is being offered. So why don't investors beat a path to their door?

This chapter will explain how investors approach a business valuation. A business valuation is the process used by investors to determine the financial worth of a closely held (private) company. For our purposes, a closely held business is an organization owned and operated by a relatively small number of owners.

The valuation process begins with an understanding of your company, following a series of generally accepted valuation techniques, such as:

- Determining which business valuations are most applicable to your company
- Analyzing the company's financial statements and pro-forma statements
- Assessing comparative and market statistical information
- Determining appropriate discounts and premiums
- Calculating current company valuation

13.2 Value vs. Price

"Some men know the price of everything and the value of nothing." —Oscar Wilde

The "**value**" of your startup is an economic theoretical concept. It is an estimate of the likely price at a given point in time. "**Price**" is the precise amount of money asked in exchange for something. In your case, you are asking Angels to invest X amount of money into your startup company in exchange for Y percentage of ownership (based on shares).

Since traditional company values are based on objective measures, such as gross revenues, net profits, cash flows, net assets, increased sales, etc., valuing a pre-revenue company is highly subjective. Complicating matters is the known "founder's syndrome," which is overvaluing your baby. It all comes down to deciding if you will be satisfied with a slice of a rapidly expanding pie (Figure 13.1).

13.3 Pre-Seed Financing

"Price is what you pay. Value is what you get."
—Donald Trump

The majority of founders contribute personal funds, along with "sweat equity," to their ventures at the pre-seed round. Founders "finance" their

Value vs Price	
Value	**Price**
■ Dynamic figure	■ Static figure
■ Nothing precise	■ Precise number
■ Subjective	■ Objective
■ Based on fundamentals	■ Based on negotiations
■ Based on economic benefits	■ Based on intrinsic benefits
■ Predictability, stability	■ If can't get champagne, get wine

Figure 13.1   Value vs. price—Do not become infected by the "founder's syndrome" by overvaluing your creation.

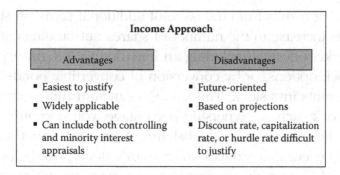

Figure 13.2   Income approach—Value determined by forecasting future earnings.

companies in one of three ways: (1) sweat equity, (2) bootstrapping, or (3) personal funds, as seen in Figure 13.2.

Sweat equity is a term of art, and is meant to represent the combined value of the knowledge, time, and effort that a founder puts into a new venture. In addition, founders also frequently engage in "bootstrapping." **Bootstrapping** is finding ways to postpone external financing or funding through creativity, ingenuity, thriftiness, cost cutting, or by other resourceful means.

13.4 What Is Your Company Worth at StartUp (Seed Round)?

"The secret of getting ahead is getting started." —Mark Twain

At the time of the seed round investment, the "initial pre-revenue valuation" represents different things to each party. To the founders, it represents the culmination of all their efforts to start the company. To the investors it provides the best early measure of their possible **return on investment (ROI)**. ROI is the concept of an investment of some resource yielding a benefit to the investor. A high ROI means the investment gains compare favorably to investment cost. In purely economic terms, it is one way of considering profits in relation to capital invested.[1]

13.4.1 Initial Valuation at Seed Round

"If you want to know the value of money, try borrowing some."

At this stage, the founders are trying to maximize the initial valuation because it will severely impact their stock dilution. **Stock dilution** is a general term that results from the issue of additional common shares by a company. This increase in the number of shares outstanding can result from a primary market offering (including an initial public offering), employees exercising stock options, or by conversion of convertible bonds, preferred shares, or warrants into stock. This dilution can shift fundamental positions of the stock such as ownership percentage, voting control, earnings per share, or the value of individual shares. A broader definition specifies dilution as any event that reduces an investor's stock price below the initial purchase price.[2] Conversely, the investors are trying to minimize the initial valuation because it represents their percentage of ownership in exchange for their invested capital. These are conflicting requirements.

13.4.2 Valuation at Series "A" Round

"It takes two to see one." —M. Szycher

Series A rounds are typically undertaken by Angels. (For more on this, see Chapter 15.) Investing in seed and startup companies is high risk. Consider that less than 10% of startups provide a reasonable return to their investors. For this reason, Angels seek only investments in companies with a high potential for quick market entry, with innovative products.

There are no good mathematical ways to calculate valuation in the absence of revenues. Therefore, Angels must rely on "qualitative" measures to assess pre-revenue valuations. Table 13.1 provides a list of the important factors Angels use in valuing their investments.

Table 13.1 Qualitative Factors Determining Valuation of Pre-Revenue Organizations

Factors	Weighted Score (%)
Management team ("Bet on jockey, not horse")	40
Opportunity ("Market pain")	20
Innovation ("Disruptive, breakthrough, groundbreaking")	20
Competition, intellectual property protection	10
Business model (marketing, sales, distribution channels)	5
Current state of company development	5

Angels target 5x to 10x ROI cash-on-cash return on their investment in 4 to 8 years, which yields an internal rate of return range (IRR) of between 25 and 75%. The **internal rate of return (IRR)** or **economic rate of return (ERR)** is a rate of return used in capital budgeting to measure and compare the profitability of investments. It is also called the discounted cash flow rate of return (DCFROR) or the rate of return (ROR).

The term *internal* refers to the fact that its calculation does not incorporate *external* factors (e.g., the interest rate or inflation rate).[3] Internal factors include:

1. Company history and its current operating performance
2. Company's projected operations and future earning capacity
3. Company's financial position and capital structure
4. Current or anticipated rate of growth in sales and profits

In the final analysis, in Series A, the actual value/price you get for your pre-revenue company depends entirely on **YOU**. It depends on what YOU negotiate. As Chester L. Karrass, the guru of negotiations famously advertises: "You don't get what you deserve, you get what you negotiate." Figure 13.3 summarizes some of the important differences between value and price.

13.5 Valuation Approaches—Companies with Revenues

"It's better to be roughly right than to be precisely wrong."
—J.M. Keynes

Before considering the various methods or procedures employed to determine value of a small company **with revenues**, we need to explore the

---

DFR method

[aka discounted cash flow (DCF)]

– Discounted Future Returns:

$$PV = \sum_{t=0}^{n} [CF_n / (1 + r)^n] + \text{Terminal Value}_t / (1 + r)^t$$

Terminal value is the present value for all cash flows (CF) in perpetuity

Present value (PV) is determined as the sum of the present value plus the present value for all years beyond the forecast period.

Figure 13.3   DFR method—Value determined by future returns, including a terminal value.

Table 13.2 Valuation Methods for Companies with Revenues	
Approach	*Based On*
Income	Present and future earnings capacity
Market	Comparable companies valuations
Asset based or cost	Current market value

three general approaches under which each of the methods fall. Table 13.2 summarizes the three accepted methods.

We will discuss each of these approaches in the subsequent paragraphs.

13.5.1 Income Approach

This determines the value based on how much the business is forecasted to earn in the future. This approach is based on the theory that fair market value is the present value of all future benefits. Fair market value is the amount at which the property would change hands between a willing buyer and a willing seller, when the buyer is not under compulsion to buy and the seller is not under compulsion to sell, and both parties have a reasonable knowledge of relevant facts.[4] Figure 13.4 illustrates the pros and cons of the income approach.

Capitalized Returns Method

This method determines value by dividing a single return (income) amount by a capitalization rate in the following formula:

Return (income) = Investment Value
Capitalization rate

Example: Calculate investment value of $150,000 at a capitalization rate (discount rate; hurdle rate) of 15%

$150,000 = $1,000,000
15%

Note: this method tends to be appropriate when the estimated future return is expected to be consistent with a normal, predictable growth rate.

Figure 13.4 Capitalized returns method—Value determined by dividing economic benefits by a capitalization rate.

Mathematically, the income approach involves computing future benefits and discounting them to their present value at a rate that reflects the inherent risks associated with the operations. The primary three include:

1. Discounted future returns method
2. Capitalized return method
3. Excess earnings method

13.5.2 Discounted Future Returns Method

Proper use of this method assumes the appraiser can estimate future returns with a reasonable degree of accuracy, and there is a reasonable probability that those returns will continue. The value determined under an income approach is based on discount or capitalization rates using public market data or internally generated hurdle rate. Hurdle rate is the minimum ROR (interest rate) on a project or investment required by an investor in order to compensate for risk. Thus, from an investment standpoint, the riskier the project, the higher the hurdle rate should be.[5]

Figure 13.5 shows Present Value (PV) multiple cash flows in multiple time periods are discounted, and it is necessary to sum them as in the figure.

13.5.3 Capitalized Returns Method

The capitalization of earnings valuation method is a method within the income approach to value whereby economic benefits for a representative single year are converted to value through division by a capitalization rate.

How Do Entrepreneurs Raise Money over time?

- Sweat Equity (Bootlegging)
- Savings, 2nd Mortgage, etc.
- SBIR (government grants)
- Friends, Family, Fools (3 Fs)
- Banks (debt)
- Angel groups
- Venture capitalist investments
- IPO (initial public offering)

Figure 13.5 How do entrepreneurs raise money—The possible sources of capital available to entrepreneurs.

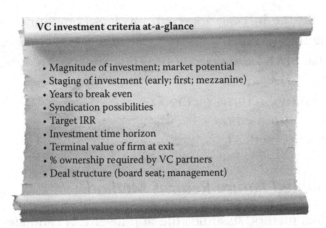

VC investment criteria at-a-glance

- Magnitude of investment; market potential
- Staging of investment (early; first; mezzanine)
- Years to break even
- Syndication possibilities
- Target IRR
- Investment time horizon
- Terminal value of firm at exit
- % ownership required by VC partners
- Deal structure (board seat; management)

Figure 13.6   VC investment criteria—How do venture capitalists think?

The valuation concept is a company's value as established primarily by the income it can be expected to earn on an ongoing annual basis, in relationship to a capitalization rate measuring ROR, investment risk, and potential earnings growth (Figure 13.6).[6]

13.5.3.1 Assumptions

Fair Value Standard of Value. "Standard of value," as it is generally defined for business valuation purposes, is the fundamental way in which the value of a business or ownership holding will be established, based on the purpose of the valuation. For the subject valuation, the standard of value being employed is "fair value," which is an opinion as to what is fair from a financial point of view as defined by law and precedent.

Net Income. "Earnings" for the method have been defined as "Net Income." The value used in the method is the one-year extension of a straight-line trend based on five historical years rounded to the nearest $1,000.

Net Income Capitalization Rate. A capitalization rate is a rate of return divisor used for converting an earnings value into an investment value. The 25.0% capitalization rate used here is a 28.0% ROR reduced by 3.0% for expected future earnings growth. Reducing the capitalization rate for the expected future earnings growth is mathematically equivalent to adjusting the earnings value for growth in perpetuity.

Discount for Lack of Marketability of 25.0%. A discount for lack of marketability is a recognized business valuation concept and, as defined in the *International Glossary of Business Valuation Terms*, is "an amount or percentage deducted from the value of an ownership interest to reflect the relative absence of marketability." The discount applied here is such a reduction in value applicable to the subject valuation.

Discount for Lack of Control of 15.0%. A discount for lack of control is a recognized business valuation concept and, as defined in the *International Glossary of Business Valuation Terms*, is "an amount or percentage deducted from the pro rata share of value of 100% of an equity interest in a business to reflect the absence of some or all of the powers of control." The discount applied here is such a reduction in value applicable to the subject holding.

13.6 Market Approach

The market approach determines the proposed value based on the price for which similar businesses are being sold. The foundation of this approach is the principle of substitution, which states, "a prudent buyer will pay no more for a property than it would cost to acquire a substitute property with the same utility." The three primary methods are:

1. Comparable sales method—actual transactions
2. Guidelines companies method—public company data
3. Industry or broker rules of thumb

13.7 Asset-Based or Cost Approach

This determines the proposed value based on what the assets in that business are worth. This approach separately values each asset and liability in the business. It does not value the unidentifiable, intangible attributes of the enterprise. The two primary methods are:

1. Adjusted book value method
2. Liquidation value method (assumes that operations will be discontinued)

13.8 Venture Capital Valuation

"Life is a dogsled team. If you ain't the lead dog, the scenery never changes." —Lewis Grizzard

If you are seeking venture capital (VC) financing, it is because you "have arrived." VC financing is the last step in the private equity market. After VC,

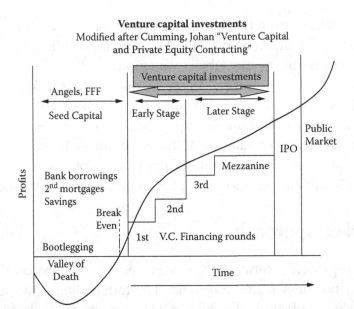

Figure 13.7   Venture capital investments—VC financial rounds start at the early stage of development.

many companies opt to go public. The entire process can be summarized as shown in Figure 13.7.

A VC fund is a financial intermediary, collecting money from investors and investing the money in certain companies on behalf of the investors. The VCs invest only in private companies that meet stringent financial requirements. After the initial investment, the VC actively monitors and helps the management of the portfolio firms. The VC mainly focuses on maximizing financial return by exiting through a sale or an initial public offering (IPO). Last, the VC invests to fund internal growth of companies (organic growth), rather than helping firms grow through acquisitions.[7]

VC is a subset of the larger private equity field, and refers to institutional investments in early-stage, high-potential, privately held growth companies. Institutional refers to investors that are not investing their own capital (like the 3Fs or Angels), but instead invest moneys obtained from pension funds, endowments, corporations, institutions, and very wealthy individuals.

VC clients only are paid when there is a liquidity event, such as a company sale or an IPO. Liquidity events are popularly known as "exits." Liquidity events are infrequent; thus, VC investments are very risky. VC is very risky because the equity investment is made in an immature company,

Target Returns for VCs

	5x	6x	7x	8x	9x
4	50%	57%	64%	68%	73%
5	38%	43%	48%	52%	55%
6	31%	35%	38%	41%	44%
7	26%	29%	32%	35%	37%
8	22%	25%	28%	30%	32%

Years to exit

IRR between 25–75%

Figure 13.8   Target returns for VCs—VCs set internal rates of return as the measuring stick for investments.

whose stock is illiquid (private) and whose share value is theoretical until the company goes public or is acquired.

VC firms must be very selective in their investments. Statistics show that of every 100 business plans evaluated by VCs, only 1 or 2 end up getting funded. Sadly, most VC-backed firms fail. Therefore, VC investments are very risky. To stay in business, VCs must ensure that their rare successes generate enough profit to compensate for their frequent failures. In an effort to lower risks, VCs utilize the investment criteria shown in Figure 13.8.

What firms are most likely to meet the VC investment criteria? VCs typically invest in companies starting at the early stage of development as shown in Figure 13.9.

IRR Required at Stages

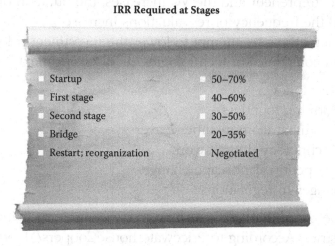

- Startup — 50–70%
- First stage — 40–60%
- Second stage — 30–50%
- Bridge — 20–35%
- Restart; reorganization — Negotiated

Figure 13.9   IRR required at stages—IRR rates decrease as the perceived risks are lowered.

As the figure indicates, VC financing rounds start at the "early" stage. The **early stage** is after a company has expended initial capital in development and market testing and is now ready to begin full-scale operations and sales. The company is producing and shipping and has growing inventories and accounts receivable. In the **later stage**, the company is breaking even or profitable and requires funds for plant expansion, full-scale marketing, and working capital. The **mezzanine** is funding provided within 6 months to 1 year of going public. Funds are to be repaid out of IPO proceeds. An **IPO** is when the company is ready to be listed in one of the major exchanges. The **public market** (MBO/acquisition) stage occurs if executives wish to purchase an independent company or a division or product line of their investor, thus creating a new independent firm, or the firm sells the business to another company.

13.8.1 Staged Capital Infusions

"Pick battles big enough to matter, small enough to win."

Rather than providing the entrepreneur all the necessary capital up front, VCs typically provide funding at discrete stages (milestones driven) over time. At the end of each stage, prospects of the firm are reevaluated. If the VC discovers some negative information, it keeps the option to abandon the project.

Staged capital infusion keeps the entrepreneur on a "short leash" and reduces his or her incentives to use the firm's capital for his or her personal benefit and at the expense of the VCs. As the potential conflict of interest between the entrepreneur and the VC increases, the duration of funding decreases and the frequency of reevaluations increases.

Staged capital infusions attempt to lower the investment risks associated with unproven companies or technologies, as discussed next:

■ Run out of money
■ Non-performing founders
■ Key management team member leaves
■ Product/technology doesn't work
■ Market/competitive landscape changes
■ Major changes in regulations

It should be noted that VC firms tend to concentrate their investments in specific industries. According to PricewaterhouseCoopers/Money Tree™, the following investments were made in 2012:

Industry	2012 Investment Total	Rank
Software	$ 1,797,910,800	1
Media and entertainment	$ 468,632,100	2
Biotechnology	$ 455,715,600	3
IT services	$ 298,444,700	4
Industrial/energy	$ 252,263,700	5
Medical devices & equipment	$ 224,805,300	6
Consumer products & services	$ 221,649,000	7
Telecommunications	$ 143,281,700	8
Semiconductors	$ 83,385,800	9
Healthcare services	$ 70,584,900	10

Examples of companies that received staged VC infusions in their early development include Apple, Federal Express, Microsoft, Genentech, and Google.

13.8.2 *Venture Capital Metrics*

"If it's stupid but works, it isn't stupid."

At the time of first investment, company valuation (pre-money) is the basic metric for calculating the theoretical ROI. Entrepreneurs need to understand that ROI is based on the potential increase in the valuation of shares received by VCs in exchange for their capital investment. Reaching agreement between entrepreneurs and VCs depends on successful agreement regarding pre-investment valuation.

Many VCs target 5x to 10x ROI (cash-on-cash ROI on the investment) in four to five years, which yields an IRR of between 25 and 75%. The **IRR** is a tool useful in corporate and personal finance for comparing investment opportunities. The goal of IRR finance is to optimize your investment funds within a defined time period to obtain the best return among competing investment opportunities.

> **Venture Capital Metrics**
>
> ■ Given a $ 1.0 m VC investment, the target IRR (40%) and the investment time horizon (5 years), calculate the future value of the VC investment
>
> ■ $FV = PV(1+i)^n$
>
> ■ i = target IRR (40%)
>
> ■ n = time horizon to exit (5 years)
>
> ■ $FV = \$ 1.0m(1+0.4)^5 = \$ 5.38\ m$

Figure 13.10   Venture capital metrics—A model calculation.

13.8.3 The Investment "Hurdle Rate"

"I hate to lose more than I love to win."

In business and investment, the minimum acceptable rate of return (MARR) or **hurdle rate** is the minimum IRR on a project a manager or company is willing to accept before starting a project, given its risk and the opportunity cost of forgoing other projects.[8] A synonym seen in many contexts is **minimum attractive rate of return**.[9] When an investment is being evaluated, it must first go through a preliminary analysis in order to determine whether it has the desired net present value using the hurdle rate as the discount rate.[10] The hurdle rate (measured as IRR) varies substantially depending on the stage of the firm at the time of investment, as shown in Figure 13.10.

Figure 13.10 presents a calculated example of the necessary future value of an investment of $1 million at the end of 5 years, assuming an IRR of 40%. Under this scenario, the investors will expect a return of $5.8 million in exchange for their initial investment. The "hurdle rate" is therefore 40%.

13.8.4 Top Ten Lies VCs Tell

"Don't pee on my leg and tell me it is raining." —Judge Judith Scheindlin

Among the most frustrating "games" that VCs play with applicants is their tendency to keep you in limbo regarding your proposal. Even after a funding presentation, this waiting game may go on for weeks, or months, if you do not push them for a definitive answer.

You feel like saying, "I can take a yes, and I can take a no, but what I cannot take is a maybe." The warning signs of a VC kiss-off are the following answers:

1. We are in the middle of several major closings.
2. It was a good presentation; I will talk with my other partners.
3. We have some major concerns about your technology.
4. Fix these issues and come back to us when they are solved.
5. The environment for funding this technology is poor at the moment.
6. Your intellectual property portfolio looks weak.
7. We need to syndicate this offering.
8. Have you been shopping this offering? For how long?
9. We are concerned about your barriers to entry.
10. You have several weaknesses in your team. You need to agree to restructure your team.

13.9 Vulture Capitalist

"A firm is known by the company it keeps."

Vulture capitalist is a derogatory term used by critics of the VC establishment. Officially, it means an investor who used deceptive clauses and terms to gain ownership of a company. A vulture capitalist focuses on investments to show an early profit and then exit quickly, trigger a takeover clause, or gain control of the company and then sell valuable parts for their profit, but in the process they may destroy the company.

True "vulture" capitalists exert their influence through onerous terms and restrictive covenants such as:

- Board seats and majority stock ownership (minimum of 51%)
- Drag-along and tag-along rights (force sale of company)
- Liquidation and dividend preferences
- Non-competition clauses
- Weighted and full ratchets (use down-round share price)
- Reserved matters (veto or approval):
  - Any sale, acquisition, merger, liquidation
  - Budget approvals
  - Executive removal/appointment
  - Strategic or business plan changes

- Anti-dilution provisions
- Right of first refusal on new share offerings
- Excessive management fees

13.10 Last Words of Advice

"Speak the truth, but shut up immediately after."

Rookie entrepreneurs have a tendency to exaggerate their technology and their accomplishments. While this may be somewhat acceptable with the 3Fs and even with Angels ("puffery"), it is a deal killer with VCs.

Most VCs are highly experienced professionals, who "have been there… done that." Following is a short list of "entrepreneur spin" that could short-circuit your plans:

- Our sales projections are super-conservative (we will gain 10% of the market in a year).
- The market potential for this technology is at least $15 billion annually.
- We don't have any competitors.
- We are about to be granted FDA approval.
- We will enjoy the first-mover advantage for several years.
- Johnson & Johnson is too slow to be a threat to us.
- We only need 1% of this market to break even.
- Our patents create an insurmountable barrier to entry.
- A number of VCs have shown great interest in funding us.
- Nobel prize winners will join us as soon as we get funded.
- The market is desperately awaiting our technology.
- This will be the last time we seek funding.

13.10.1 *How to Maximize Your Company's Value*

"Value is in the eyes of the beholder."

This last section deals with the art of positioning and communicating your story. You must develop your optimal communication platform. Investors cannot get all the information they seek from a business plan; they also need the "touch and feel" and emotions of management.

The expression "rubbing shoulders with investors" connotes the feeling of trust, credibility, and confidence building that only the human touch can provide. Therefore, a basic understanding of your maximal "intrinsic value" is in order, as shown next:

- Understand that investors' central concern is their ROI risk and how quickly they can exit.
- Companies are attractive only if they can create economic value for their shareholders.
- Founders, in turn, must create value by investing capital at rates of return exceeding their cost of capital.
- Make the strongest case for your company's current value and its future intrinsic value by being trustworthy.

References

1. http://en.wikipedia.org/wiki/Return_on_investment
2. http://en.wikipedia.org/wiki/Stock_dilution
3. http://en.wikipedia.org/wiki/Internal_rate_of_return
4. IRS Revenue Ruling 59-60 (Appendix G).
5. http://www.investopedia.com/terms/h/hurdlerate.asp
6. http://www.valuationcalcs.com/CapEarningsFolder/SampleReport.asp
7. Venture Capital Funds. http://home.ku.edu.tr/~cdemiroglu/Teaching/MFIN514new/VC.funds.ppt
8. Lang, H.J., & Merino, D.N. *The Selection Process for Capital Projects.* New York: John Wiley & Sons, 1993.
9. Ross, S.A., Westerfield, R.W., & Jaffe, J. *Corporate Finance*, 7th ed. McGraw Hill International Edition, 2005.
10. http://en.wikipedia.org/wiki/Minimum_acceptable_rate_of_return

Chapter 14

Writing a Winning
Business Plan

14.1 Introduction

A business plan is a selling document that conveys your irrepressible excitement and vision to potential investors and stakeholders. The creation of your business plan is the first step on your path to financial success. The business plan is your key to achieving the entrepreneurial goals of the 4 Cs: (1) cash, (2) control, (3) creativity, and (4) challenge.

A business plan is your most crucial business document. Without a persuasive business plan, no one will seriously consider investing in your startup venture; and without investors, it is unlikely you will ever get off the ground.

As you go through this chapter, remember that the greatest beneficiary of the business plan is not your investor, customer, or banker—it is you. Whether your enterprise is a startup, small, medium-sized, or rapidly growing business, develop a winning business plan to enable you to:

1. Persuade financial investors.
2. Assess if your business ideas will work.
3. Outline each area of the business and maximize your resources.
4. Establish financial and technical milestones.
5. Define your target market and set profitable directions.
6. Secure additional funding or loans for continuing growth.

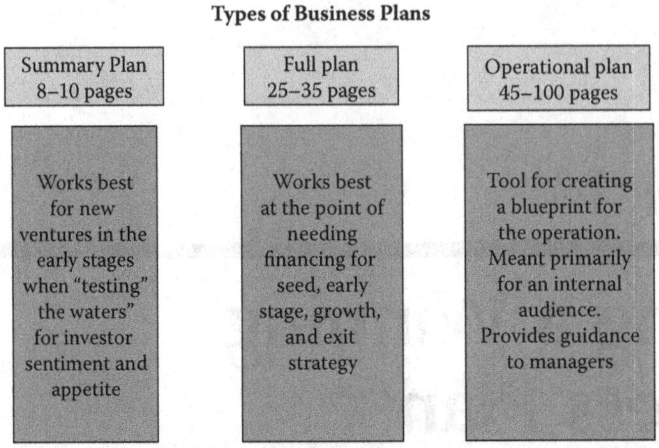

Types of Business Plans

Summary Plan 8–10 pages	Full plan 25–35 pages	Operational plan 45–100 pages
Works best for new ventures in the early stages when "testing" the waters" for investor sentiment and appetite	Works best at the point of needing financing for seed, early stage, growth, and exit strategy	Tool for creating a blueprint for the operation. Meant primarily for an internal audience. Provides guidance to managers

Figure 14.1   Types of business plans—The three types of business plans.

7. Determine your long-term financial needs.
8. Attract and retain top-level personnel.
9. Help monitor the progress of your business over time.
10. Devise contingency plans and avoid obstacles.

In general, there are three types of business plans, as shown in Figure 14.1.

Your plan will be read by two primary audiences: investors and other stakeholders (for the Summary and Full Plans), or your own employees (for the Operational Plan). Thus, your business plan must make the case that it will serve the interests of investors and stakeholders. The key is to include facts generated through a properly conducted feasibility analysis. A business plan rings hollow if it is based strictly on what founders project or envision will happen. This is summarized in Figure 14.2.

14.2 Need for a Winning Business Plan

"The only good business plan is the one that gets funded."

Perhaps you are going through the motions of writing a business plan because both your attorney and your accountant told you that investors require a well-reasoned business plan. Most entrepreneurs view the business

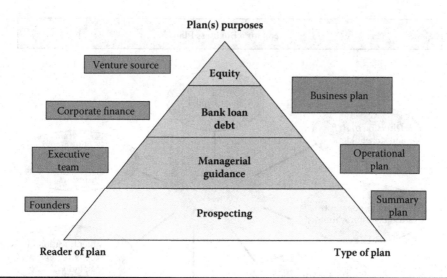

Figure 14.2   Plan purposes—Your plan must emphasize the interests of investors and stakeholders.

plan as a chore, secondary to technological achievement and innovation. On the contrary, a business plan is worth your considerable investment of time, effort, and energy.

A business plan is your only chance to demonstrate that you know what you are doing and can be trusted with the money you are seeking. This is summarized in Figure 14.3.

Your full business plan should be structured to answer the following investor questions:

What ?	Business are you in?
How?	Much money do you really need?
Who?	Is in the deal?
What?	Is unique about this deal?
Why?	Should I invest now?

Do not make the plan too long.

   "If an investor is interested, they'll ask for more details." —Eugene Kleiner, Venture Capitalist

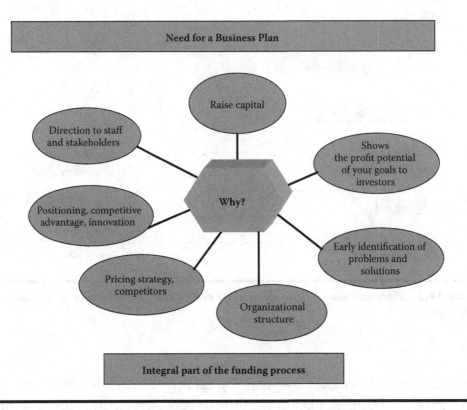

Figure 14.3 Need for a business plan—Demonstrate to investors that you can be trusted with their funds.

14.3 The Essential Components

"If you don't know where you are going, any road will take you there."

While you may be tempted to be creative with your business plan, at present, most business plans follow an accepted and pre-described format. It forces the entrepreneur to set objectives, define budgets, engage staff, anticipate problems, and identify how the innovation will solve market needs. It attempts to prevent the entrepreneur from catching the well-known Christopher Columbus syndrome:

■ When he left, didn't know where he was going.
■ When he got there, didn't know he had arrived.
■ When he returned, didn't know where he had been.

According to the Federal Development Insurance Corporation (FDIC),[1] every business plan should include four essential components:

1. Overview of the business: Describes the business, including its products and services.
2. Operations and management plan: Describes how you will manage the core processes of your business, including use of human resources.
3. Marketing plan: Describes the target market for your product and explains how you will reach that market.
4. Financial management plan: Details the costs associated with operating your business and explains how you will pay for those costs, including the amount of financing you may need.

14.3.1 Common Parts of a Good Business Plan

"The only place success comes before work is in the dictionary."
—Vidal Sassoon

Business plans must help investors understand and gain confidence in how you will meet your customers' needs. Your language should give the impression that you are thoughtful, knowledgeable, and trustworthy. Your statements should be verifiable by credible references. Do not overstate your enthusiasm with exuberant phrases such as "we have no competition," "unmatched," "terrific," etc.

The common parts of a good business plan are shown in Table 14.1.

14.4 Risk Is a Four-Letter Word

"Your business plan is nothing more than a reasoned list of risk-mitigating milestones." —M. Szycher

Your best chance of being funded is to *persuade* investors that you are mitigating all known risks, and thus protecting their investment. Notice that I said "persuade," not "convince." **Convincing** aims for investors to accept the problem as you have presented; **persuading** aims to move investors to act on your proposal and invest in your company. This is shown graphically in Figure 14.4.

Table 14.1   Summary of Common Parts Needed in a Business Plan

I	Cover sheet	Company information
		Name, title, and phone number of corporate officers
		Copy number of the plan
II	Executive Summary	Critical because many investors only read the summary
		Market opportunity
		Capital requirements
		Competitive advantage
III	Company Description Business Concept	Shows evidence that your product or service is viable and capable of fulfilling an unmet market need
		What you sell; benefits, features, description
		Captures and highlights the value proposition in your product or service offerings
		SWOT analysis
IV	Marketing Plan Market Analysis Industry Analysis and Trends Competition Sales Strategy	Defines your target market
		Segments your customers
		Market growth and forecast
		Projects your market share and trends
		Positions your products and services
V	Management Team and Organization	Investors bet on the jockey (management), not the horse (technology)
		Highlights the track record of the key members; you may also offer details about key employees including qualifications, experiences, or outstanding skills, which could add a competitive edge to the image of the business
		Management gaps and personnel plan

Table 14.1 Summary of Common Parts Needed in a Business Plan (Continued)

VI	Technology and Innovation Risk Assessment Operations and Management Plan	Intellectual property portfolio
		Governmental regulations to be followed and their impact
		Organizational structure of the company
		Provides a basis for projected operating expenses
		Because these statements are heavily scrutinized by investors, the organizational structure has to be well-defined and realistic within the parameters of the business
		Expense and capital requirements to support the organizational structure
		Provides a basis for identifying personnel expenses, overhead expenses, and costs of products/services sold; these expenses/costs can then be matched with capital requirements
VII	Milestones and Exit Plan	Funding strategy
		Risk evaluation (economic, financial, technological)
		Timetable of expected business outcomes (start/end dates)
		How investors will get their money back, with interest
VIII	Financial Plan	Clearly defines what successful outcomes entail; your plan isn't merely a prediction; it implies a commitment to making the targeted results happen and establishes milestones for gauging the organizational progress
		Provides a vital feedback-and-control tool for investors and staff
		Variances from projections provide early warnings of problems
		When variances occur, the plan can provide a framework for determining the financial impact and the effects of various corrective actions
		Three-year income, balance sheet, and cash flow projections
		Break-even analysis
		Sources and uses of funds
		For the first three years, all you have to worry about is cash flow

(Continued)

Table 14.1 Summary of Common Parts Needed in a Business Plan (Continued)

		If your business is new, and pre-revenue, you may stop here
		Monthly financials are necessary for the first three years
		Annual projections will suffice for the fourth and fifth years
		If your business is established, you will include actual statements
IX	Supporting Documents	Leases, mortgages, business agreements, contracts, etc.

Investors make their decisions based on emotion (convincing), and justify them with logic (persuasion). Turn your business plan into an irresistible offer. Your irresistible offer[2] should consist of four elements:

- High return of investment (ROI)
- Recognizable market need
- Credible risk mitigation strategy
- Reasonable exit strategy

14.5 Management Team That Can Execute

"The best companies solve 100 problems at a time; the worst solve one problem 100 times." —Randy Pennington in *Results Rule!*

Perhaps more than any other factor, a competent management that can execute is the most important ingredient in business success. Persuade investors

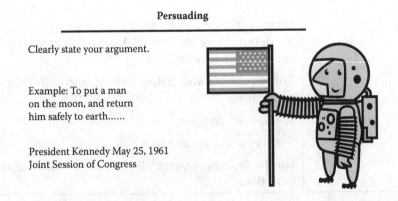

Persuading

Clearly state your argument.

Example: To put a man on the moon, and return him safely to earth......

President Kennedy May 25, 1961
Joint Session of Congress

Figure 14.4 Persuading—Persuade investors that you are mitigating all known risks.

that you have assembled a management team capable of executing the mission, have defined accountability, and are focused on performance and results. A superb management team will be capable of solving unforeseen problems, managing the dynamics of change, and overcoming the inevitable crises that are sure to come.

Before submitting your plan to investors, conduct your own assessment of your management team. Critically evaluate each executive, and highlight the following traits:

- **Experience.** Industry experience, solid management background, trustworthy.
- **Realism.** The team fully understands the many needs and challenges of the business.
- **Flexibility.** Aware of the team's limitations, recognizing the need for planning and accountability.
- **Ability to get along.** The team recognizes its ability to motivate, reward, and punish.

14.5.1 Competent Team

"By working faithfully eight hours a day you may eventually get to be boss and work twelve hours a day." —Robert Frost

According to Patrick Lencioni,[3] teamwork is a unique, powerful advantage unmatched by technology, strategy, or finance. He lists the five dysfunctions of a team as:

1. Inattention to results
2. Avoidance of accountability
3. Lack of commitment
4. Fear of conflict
5. Absence of trust

To mitigate risks, you must persuade investors that you have assembled a real team, bereft of the five known dysfunctions. You have successfully assembled a group of entrepreneurs that have common goals and share both rewards and accountability for executing your stated business goals.

14.6 How Innovative Is Your Invention?

"The harder I work, the luckier I get." —Thomas Edison

Did you know that Thomas Edison *did not* invent the light bulb? Twenty-two prior inventors worked on incandescent electric light bulbs before Edison. However, Edison best understood the market need, and realized that a commercially viable and longer lasting light bulb was required for the technology to go "mainstream." Therefore, he and his team created the product, and the rest is history.

Edison also relied on investors. He persuaded skeptical investors that his team was capable of executing. The moral of the story is that if the idea is so good and the market need so intense, many other people have also seen the opportunity, either before you or at the same time. You must persuade investors that you and your team will be the winner that will open the market, tweak the invention based on customer feedback, and scale the business to commercial success. At this point, you should be prepared to answer the three "why" questions of investors:

1. Why you?
2. Why me?
3. Why now?

14.7 Ignore Naysayers (and Prove Them Wrong)

"I stand on the shoulders of giants." —Sir Isaac Newton

Do not be deterred by critics. Of course, your innovation is unproven in the marketplace. If it were proven, it would not be an innovation, right? Even Einstein relied on previous attempts to develop his innovative theory of relativity.

And of course, others have tried it before, but failed. If they had not failed, everyone would be using it already. Right again? You must persuade your potential investors that you have found the answer to the problem.

Even the smartest people fail to envision the future. Following are famous examples, or what we will call the Hall of Shame:

■ "Airplanes are interesting toys but of no military value." —Marechal Ferdinand Foch, Professor of Strategy, Ecole Superieure de Guerre.

- "I think there is a world market for maybe five computers." —T.J. Watson, Chairman, IBM, 1943.
- "There is no reason for any individual to have a computer in his home." —Ken Olson, founder and Chairman, Digital Equipment Corporation (DEC), 1977.
- "But what ... is it good for?"—Engineer at Advanced Computing Systems Division of IBM, commenting on the microchip, 1968.
- "Computers in the future may weigh no more than 1.5 tons." —*Popular Mechanics*, forecasting science trends, 1949.
- "A cookie store is a bad idea. Market research reports say America likes crispy cookies, not soft and chewy like yours." —Response to Debbi Fields' inquiry on her business potential.
- "Drill for oil? You mean drill into the ground and try to find oil? You're crazy." —Drillers when Edwin L. Drake tried to enlist them to drill for oil in 1859.
- "This 'telephone' has too many shortcomings as a means of communication. It is inherently of no value to us." —Western Union internal message in response to Bell's proposal, 1876.
- "We don't tell you how to coach, so don't tell us how to make shoes." —Large sporting goods shoe manufacturer to Bill Bowerman, co-founder of NIKE, Inc.
- "The concept is interesting and well formed, but to get better than a C grade, the concept must be feasible." —Yale University professor to Fred Smith's paper for overnight delivery service. Smith subsequently found Federal Express Corporation.
- "Who the hell wants to hear actors talk?" —H.M. Warner, Warner Brothers, 1927.
- "The television will never achieve popularity; it takes place in a semi-darkened room and demands continuous attention." —Harvard Professor Chester L. Dawes, 1940.
- "What's all this computer nonsense you're trying to bring into medicine? I've got no confidence at all in computers, and I want nothing to do with them." —Distinguished medical professor in England to Dr. John Alfred Powell, eventual developer of the revolutionary CT scanner.

The laughs are on me, my fellow entrepreneurs.

14.8 Milestones Make Your Plan Real and Believable

"Completing your business plan would be a milestone in itself."

Your milestones table is one of the most important in your business plan.[4] It sets the plan into practical, concrete terms, with real budgets, deadlines, and management responsibilities. It helps you focus as you are writing your business plan, and plan-vs.-actual management analysis helps you implement your plan as you grow your business.

Milestones are points or deliverables in your plan that help you to know that you are making progress and moving in the right direction. Otherwise, without them, you really don't know if things are happening as predicted. Funding sources look at the milestones primarily for answers to the crucial questions concerned with the heart of the plan:

■ Is the business model solid and attainable?
■ Is the market large enough to sustain the business as proposed?
■ Do you have an unfair competitive advantage?
■ Are there significant barriers to entry?
■ Is key management experienced and capable?
■ Are the financial projections realistic?
■ Does the plan describe the exit strategy?

14.9 The Value of Your Plan Will Be Measured by Its Implementation Potential

"Don't develop paralysis of the analysis."

The milestones should be one of the most important sections of the entire business plan. Each marketing and sales-related program you plan should be listed in the table and explained in the accompanying text, along with relevant details. You want to cement your sales strategy with programs that make it real.

■ How is this strategy to be implemented?
■ Do you have concrete and specific plans?
■ How will implementation and success be measured?

14.10 Example of a Winning Business Plan

References

1. Building a Business Plan. http://www.fdic.gov/about/diversity/sbrp/12.ppt
2. Joyner, M. *The Irresistible Offer*. Hoboken, NJ: John Wiley & Sons, 2005.
3. Lencioni, P. The Five Dysfunctions of a Team. http://flpbs.fmhi.usf.edu/pdfs/Five%20Dysfunctions%20of%20a%20Team.pdf.
4. Barry, T. Milestones Make Your Business Plan a Real Plan. http://articles.bplans.com/writing-a-business-plan/milestones-make-your-business-plan-a-real-plan/44.

POLYMEDICA INDUSTRIES, INC.

(Advanced Polymer Technology)

BUSINESS PLAN

JANUARY 1991

CONFIDENTIAL

Document Number _____

Two Constitution Way
Woburn, MA 01801

Telephone: 617/933-2020
Telecopier: 617/933-7992

TABLE OF CONTENTS

Investment Summary...1

The Company and Its Products 3

    Mitraflex™ Wound Care Products 5
    Biostable Polyurethanes 9
    Chronosphere™ Microparticulates 12
    Chronothane™ Photopolymers 15

Polyurethanes 21

    Table I: Polymers Accepted for Use
                in Medical Applications 23
    Table II: Device/Function/Biomaterial 24
    Table III: Synthetic Polymers 25

Management 27

Patents 32

Capitalization 36

    Table IV: Issued and Outstanding Shares
                as of November 30, 1990 37
    Table V: Fully Diluted Capitalization
                after Series B Preferred
                Stock Financing 38

Capital Requirements 39

Operating Plan 44

Appendices:

    Appendix A: Financial Statements and Auditors
                    Opinion
    Appendix B: Detailed Financial Projections:
                    FYE March 31, 1992 –1996
                    Pro Forma Sales
                    Pro Forma Statements of Operations
                    Pro Forma Balance Sheets
                    Pro Forma Cash Flows
    Appendix C: Glossary of Terms

Confidential   **PolyMedica Industries, Inc.**

Confidential
Information

The information in this Business Plan is the confidential and proprietary written property of PolyMedica Industries, Inc. ("PolyMedica" or the "Company") and is being submitted to prospective investors in the Company solely for each investor's confidential use with the express understanding that, without, prior written permission of the Company, those persons will not release this document, discuss the information contained herein, or make reproductions of or use this Business Plan, for any purpose.

By accepting delivery of this Business Plan, a prospective investor agrees to promptly return to the Company this Business Plan, and any other document or information furnished, and all copies of any of the foregoing, if so requested by the Company.

THE COMPANY AND ITS PRODUCTS

COMPANY HISTORY

The Company was seed funded in November 1989, under the name of Emerging Sciences, Inc., to commercialize a proprietary family of polyurethane elastomers which can be tailored to specific products. Application of these polyurethanes includes controlled medical and cosmetic delivery systems as well as a wide range of other industrial products. Its initial round of financing was provided by Commonwealth BioVentures, Inc. in the amount of $400,000 and enabled the founders to secure a facility and prove the technology. Early success brought about cooperative projects with a number of large cosmetic and medical device joint-venture partners.

In June 1990, Emerging Sciences, Inc. secured an additional $2.4 million of financing which enabled the Company to purchase the assets of a former public company called Matrix Membranes, Inc. ("Matrix"). Matrix had developed a wound dressing product based on a complementary aromatic polyurethane technology. This acquisition broadened the technical base of the Company and brought a unique capability to tailor polyurethanes to numerous medical applications. In July 1990, the name of the Company was formally changed to PolyMedica Industries, Inc., to clearly reflect its core technologies. PolyMedica manufactures products based on aromatic or aliphatic polyurethanes and solutions or solid polymers, as well as photopolymers and heat-cured polymers.

STRATEGIC MISSION

PolyMedica is committed to leadership in advanced polymer technology and its application to growth businesses. The Company can formulate and manufacture polyurethane-based products to solve problems where no other material is suitable.

The polyurethanes which PolyMedica can synthesize have unusual physical and chemical properties which are tailored to the special needs of the medical field and other high technology industries. Our ability to customize complex polymers, and the processes to manufacture them, is unique. The Company has a broad technical base which includes the ability to manufacture raw materials. These skills position PolyMedica to take advantage of new and rapidly growing medical, personal care and industrial markets.

FACILITIES

PolyMedica currently has two combination manufacturing and laboratory facilities located at separate sites. One is in Woburn, Massachusetts (9,000 sq. ft.) and the other in Wheat Ridge, Colorado (10,000 sq. ft.). The Woburn plant will manufacture UV curable, aliphatic polymers which are formed as films and microparticulates. The Company took possession of the Woburn facility on November 1, 1990. Under the terms of a five year lease, the Woburn facility was built to PolyMedica's custom specifications. The equipment will consist of mixing tanks, film forming equipment, an irradiation chamber exposing the product to long wave length UV light, and granulators. The manufacturing process transforms the proprietary chemical formulations into thin layers of polymer which are then used as films or microparticulates. The products which will be manufactured at the Woburn plant are currently being developed with major pharmaceutical, personal care and industrial partners who are seeking to apply the technology for the products in such areas as implants, transdermal patches, vascular grafts and cosmetics.

The Wheat Ridge, Colorado facility manufactures Mitraflex™ wound dressings and derivative products. The types of equipment in place consist of chemical reactors, customized casting and polymer forming machinery, die cutting equipment, and pouching and sealing equipment. This equipment is used to prepare and transform proprietary polyurethane formulations and manufacturing processes into the various layers of the wound dressing (namely, membrane, top thin film layer, and adhesive). These three layers are laminated together to form the unique Mitraflex™ wound dressing. After the product is die cut, pouched, and boxed, it is sent to a contract sterilizer where it receives appropriate gamma radiation before being shipped to customers.

The Company uses materials requirement planning (MRP) systems for its manufacturing operations, with appropriate financial and cost accounting reporting.

PERSONNEL

The Company currently employs 24 people.

PRODUCTS

PolyMedica can design and manufacture polyurethanes as liquids, solids, films, particles and membranes. The Company's products can be broadly classified into wound care, biostable polyurethanes, microparticulates, and photopolymers. The following is a discussion of each of the Company's products.

1. MITRAFLEX™ WOUND CARE PRODUCTS

1.1 General Description of Wounds

A wound is a disruption in the anatomy or physiology of tissue. Wounds are caused either surgically or accidentally as in burns or trauma. They vary in severity from minor scratches to complicated trauma.

Wound management has developed rapidly in the past few years with the advent of new wound dressings and a better understanding of the physiology of wound healing. Implementation by the Federal Government of new regulations for medical reimbursement increases the opportunity for innovative products that speed wound healing in hospital situations. In addition, demand for wound care products has expanded in alternative sites such as nursing homes and individual patient homes.

1.2 Decubitus Ulcer Wounds

Decubitus ulcers, or pressure sores, are common complications in bedridden patients. These ulcers are caused by unremitting external pressure that impedes blood flow to surrounding tissues and results in necrosis. PolyMedica's wound dressings are especially effective in treating this common type of wound.

Approximately 70% of all pressure sores occur in elderly patients. The cost of treating decubitus ulcers is surprisingly high. However, government payment for therapy treatment that usually requires 15 inpatient days is only $6,500, or $430 per day. This funding level does not cover the room cost and medical supervision to treat this serious problem. There is no payment for hospital-acquired pressure sores and when these occur in hospitalized patients, the hospital facility must underwrite the expense of the treatment.

Pressure sore development can become a malpractice issue if the wound develops while a patient is in the care and custody of the hospital. The incidence of decubitus ulcers ranges from 1.5 to 3 million occurrences per year.

1.3 Wound Care Markets

The annual U.S. market for wound care products was $1.76 billion in 1988 and is forecasted to increase to $2.27 billion by 1992. Some of the major areas are (in $ million):

Confidential   **PolyMedica Industries, Inc.**

Year	Bandages & Dressings	Burn Care	Wound Closure
1988	$ 93.2	$46.8	$ 851.9
1989	97.0	48.7	923.4
1990	101.1	50.8	1,004.5
1991	105.7	53.1	1,091.0
1992	110.5	55.4	1,187.4

Source: Biomedical Business International

Within the context of the overall market, the newer generation of synthetic products is forecasted to grow at 21% annually. Newer dressings can accelerate wound repair and also prevent infection. The favorable trade-off between the higher cost of these synthetic wound dressings and the reduced cost of treatment and liability will produce rapid growth in the markets for the newer dressings.

CROSS SECTION OF MITRAFLEX™ WOUND DRESSING

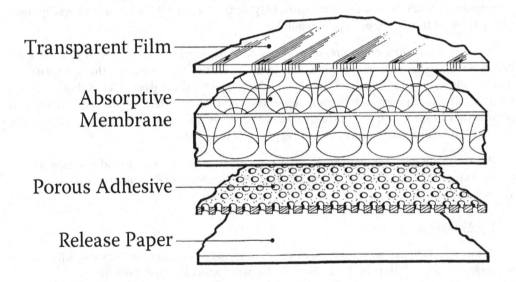

Transparent Film

Absorptive Membrane

Porous Adhesive

Release Paper

1.4 The PolyMedica Mitraflex™ Wound Dressing

PolyMedica's Mitraflex™ wound dressings represent a new generation of spyrosorbent (both breathes and absorbs), patented products affording significant advantages over the hydrocolloid (water absorbing) and thin-film dressings generally available in today's wound care market. While Mitraflex™ dressings are specifically designed to treat dermal ulcers, they also perform exceptionally well on abrasions, lacerations, donor sites, burns, and dermabrasions.

These dressings combine high exudate absorption and moisture vapor transmission capabilities to maximize wound healing through creation of an optimal moist healing environment. The unique combination of these two moisture management methods in a single product results in a low profile dressing that is conformable to the complex human skin anatomy and effectively manages wound exudate. The dressings allow visual monitoring of the exudate without removal for periods up to seven days. This is a feature of particular interest to many clinicians. Finally, the conformability of Mitraflex™ allows a placement on and consistent adhesion to body contours that no other wound dressing can protect.

The dressings are protected under four patents (see "Patents") in the name of PolyMedica's Mitrathane™ medical-grade polyurethane and are manufactured at the Wheat Ridge, Colorado facility.

PolyMedica entered into a distribution agreement dated January 1, 1990 with Calgon Vestal Laboratories, a division of Calgorl Corporation, a wholly-owned subsidiary of Merck & Co., Inc. This agreement grants Calgon the exclusive right to sell Mitraflex™ wound dressings in two specified sizes (4"×4" and 8"×8") in the U.S. and Canada through 1995. Calgon has developed a complete marketing program with literature and promotional material, and is currently marketing and selling Mitraflex™ through its 186-person sales force in North America.

As of November 1990, approximately 900,000 Mitraflex™ wound dressings have been manufactured and sold. Due to its exceptional clinical results, Mitraflex™ has been able to command a premium price at retail of $4.25 for a 4"×4" size and $8.00 for an 8"×8" size.

1.5 Other Uses of Mitraflex™

The microporous, absorbent nature of Mitraflex™ dressings can also be applied to other wounds, such as dermabrasions and donor sites. Dermabrasions occur during cosmetic abrasion of the facial dermis of patients suffering from pock marks or severe acne. Donor sites are areas of the skin that have been "shaven" of dermis to donate skin to another area of skin that has been burned.

In both cases, Mitraflex™ acts as a temporary artificial skin, which:

> Adheres gently to intact skin
> Maintains optimal fluid balance
> Matches the elasticity of normal skin
> Prevents mechanical abrasion
> Maintains temperature homeostasis
> Accelerates wound healing (compared to gauze treated wounds).

PolyMedica is well advanced in adapting the Mitraflex™ technology to other markets where these characteristics represent advancements in the state-of-the-art. These include such market segments as plastic surgery wound dressings, synthetic gauze dressings, veterinary wound dressings, pigmented wound dressings, and variable adhesive dressings. These are products that PolyMedica can produce and market on its own or in conjunction with a marketing partner.

In addition, the Company is pursuing longer term possibilities of developing wound dressings capable of delivering over-the-counter medications or ethical pharmaceuticals at the wound site through external application. While no agreement has been finalized, several large companies have expressed interest in developing a partnership with PolyMedica for entry into this high sales potential market segment.

1.6 Expanding the Markets for Mitraflex™

PolyMedica is in the process of negotiating with other distributors for its Mitraflex™ products in Europe, Japan, South America, and other international markets. The Company intends to apply some of the proceeds of future financings to increase product sales efforts (see "Capital Requirements").

2. BIOSTABLE POLYURETHANES

2.1 Definition of Biostable Polyurethanes

Polyurethanes ("PU's") are the best performing biomedical-grade elastomers. The only other suitable elastomer is the family known as silicone. Silicones are biostable, histocompatible and hemocompatible, but are mechanically weak. Such mechanical properties could be overcome by increasing the thickness of the silicone-based product. However, this is not possible or desirable in most applications.

Polyurethanes are block copolymers synthesized from isocyanates, macroglycols, and chain extenders. By proper selection of these three monomers, PU's can be made flexible, semirigid, or rigid. They have demonstrated high hemocompatibility, superior abrasion resistance, excellent mechanical properties, and unmatched flexure endurance. Due to these properties, polyurethanes are the polymer of choice for the manufacture of the most advanced medical devices.

In the past five years, research has shown that all currently available medical grade polyurethanes undergo an enzymatic degradation process known as Environmental Stress Cracking ("ESC"). ESC, as the name suggests, leads to surface microcracks in the corrosive environment of the living body. Once microcracking begins, it is only a matter of time before catastrophic failures of the medical device occur.

Scientists at PolyMedica have developed Chronoflex™, the first biostable PU elastomer that initial tests indicate may be capable of withstanding the high corrosive biological environment of the human body. Chronoflex™ is presently undergoing tests in three areas: (1) Vascular Grafts; (2) Pacemaker Lead Insulators; and (3) Mammary Prostheses.

2.2 Vascular Grafts

Vascular disease, such as atherosclerosis, is usually progressive. A fatty streak in the vessel can rapidly develop plaques that ultimately impede blood flow to the heart muscle. Many patients eventually need vascular grafts to bypass severely obstructed coronary arteries.

Confidential   **PolyMedica Industries, Inc.**

Two biomaterials currently dominate the graft field: Dacron and Teflon (PTFE). Dacron monofilaments are woven into various intricate designs. Dacron grafts are generally used in large diameter (12 to 22mm) applications. Blown PTFE grafts are generally used in the intermediate diameters (6 to 12mm).

Three companies dominate the vascular graft field: C.R. Bard (Dacron), W.L. Gore (PTFE), and Meadox (Dacron).

Dacron and PTFE grafts are acceptable for peripheral vascular surgery, shunts and aneurysm repairs where the diameter exceeds 6mm. For diameters smaller than 6mm, which is a large segment of the market, studies indicate that miscompliance results in anastomotic hyperplasia, thickened fibrin linings and kinking, with postoperative failures as high as 50% at 12 weeks.

Polyurethane grafts with improved compliance-matching to the native coronary vessels are, at present, the best alternative. However, with the specter of surface microcracking inherent in the commercially available polyurethanes, the industry has sought the development of a biostable polyurethane. PolyMedica believes it has developed such a polyurethane in its Chronoflex™ product.

To date, no organization is marketing arterial grafts suitable for the bypass of obstructed coronary arteries. Each year, there are 320,000 coronary bypass surgeries in the U.S. Surgeons still rely on the use of saphenous veins from the patient, and are vitally interested in the development of an off-the-shelf artificial vascular graft substitute. A major new market opportunity exists (in excess of $150 million per year) for the development of small bore (6mm dia. or less) vascular grafts. If PolyMedica's biostable polyurethane product can be developed for peripheral and vascular applications, the graft market should grow to $500 million annually as new sizes are introduced.

2.3 Pacemaker Lead Insulators

Cardiac pacing systems are utilized by patients with some impairment in the heart's natural electrical conduction system. Normally, the rhythmic contractions of the heart are stimulated by small electrical signals emitted by the sinoatrial node. These signals are conducted downward along nerve fibers to the four chambers of the heart (first to the atria and finally to the ventricles) which contract to perform the principal cardiac pumping function.

When the natural pacing mechanism is impaired, a pacemaker is used to remedy the problem by stimulating the heart and restoring normal rhythm. Pacemakers are implanted subcutaneously and are connected to the heart muscle by means of insulated wire leads called electrodes. The useful life of pacemakers ranges from five to ten years, but the leads are expected to survive indefinitely.

Advances in cardiac pacemakers are allowing more patients to play tennis, swim, and even jog. That is important since almost 250,000 patients received pacemakers worldwide—a quarter of them between the ages of 25 and 45—accounting for industry sales that exceed $1 billion annually. Approximately 55% of implants are performed in the U.S.

Until 1981, most lead insulators were made of silicone. However, silicone insulators were very thick so the industry began switching to polyurethane insulators. Compared to standard silicone rubber, PU elastomers offered several advantages, including higher tensile strength, tear resistance, and abrasion resistance.

These advantages prompted the introduction of PU leads with significantly reduced wall thicknesses by Medtronic, Cordis, and Intermedics. Thinner leads allowed easier surgical insertion, the introduction of multiple leads, and greater elasticity for the implanted lead.

The drawback once again is that these polyurethane leads are susceptible to microcracking. Chronoflex™ overcomes this difficulty; samples of Chronoflex™ tubing are undergoing final in vivo stability tests at several laboratories.

2.4 Mammary Implants

Over 125,000 women undergo breast augmentation and reconstruction in the U.S. every year. The total population of U.S. implants now exceeds 425,000 women. The major suppliers marketing mammary prostheses are Mentor, McGhan, Dow-Corning Wright, and Surgitek. It is estimated that the U.S. mammary prostheses market for these companies is about $75 million per year.

Most prostheses are composed of an outer wall of silicone rubber, filled with a silicone gel core. These prostheses are prone to leak silicone gel and frequently exhibit capsular contraction. The leakage introduces foreign material into the body and capsular contraction makes the breast hard, tender, and unnatural in appearance. These problems have prompted the FDA to reclassify current prostheses and move them from Class I to Class III Devices requiring a stringent PMA (Premarket Approval) pathway.

Most authorities agree that capsular contraction can be effectively eliminated by "coating" the outer wall of the mammary implant with a suitable tissue ingrowth platform. No acceptable material is currently available to manufacturers of breast prostheses that allows tissue ingrowth while remaining biostable.

PolyMedica has pioneered the development of a proprietary medical textile that can be used as a tissue ingrowth platform. Its medical textile is based on a novel forty filament fiber which is melt extruded from a Chronoflex™ elastomer. Prototypes of this textile have been manufactured by means of a sophisticated polar winding technique.

PolyMedica is currently negotiating an agreement with a major strategic partner for the development, testing, manufacture, and distribution of a series of product(s) emanating from this technology.

3. CHRONOSPHERE™ MICROPARTICULATES

3.1 Description of Microparticulates

The concept of entrapping materials within a polymeric coating is commonly called microencapsulation. These capsules all vary in size from less than one to as large as 2,000 microns. Historical examples of the use of the microencapsulation process in the pharmaceutical industry are the Spansules[R] of SmithKline, introduced in the 1950s for the sustained release of orally administered drugs. Estee Lauder introduced in the mid-1980s a liposome-based moisturizing cream which released hydrating agent overnight.

PolyMedica has developed a new technology which delivers ingredients by the process of diffusion. The diffusion process involves the entrapment of active ingredients in an inactive carrier matrix, which releases the ingredients over a prolonged period of time.

3.2 Technical Background

In the pharmaceutical and personal care industries, microencapsulation of active ingredients can:

Sustain the effectiveness of such ingredients
Permit liquids to be handled as solids
Protect reactive components until time of use
Allow the safe handling of toxic materials
Extend the shelf life of delicate materials
Overcome product incompatibilities.

In topical products, which are generally liquids or creams, microcapsules have not been used. Current soluble or pressure sensitive microcapsules will release actives in storage if they are contained within a water-based lotion or cream or will release actives upon breakage during processing. Insoluble microcapsules containing an active ingredient will not release their contents when it comes in contact with skin. PolyMedica has developed its Chronosphere™ technology to overcome these two problems.

3.3 Description of Chronospheres

Chronosphere™ polymer microparticulates represent an advanced technological system for the sustained and controlled delivery of cosmetic and pharmaceutical actives to the skin in a traditional liquid, semi-solid, or powder carrier.

Chronospheres are a matrix system in which the active ingredient is dissolved or dispersed into the liquid prepolymer. The solution occurs at a molecular level and insures a uniform concentration of active ingredient throughout the molecular particle. Since the composition of the polymer can be altered, its hydrophilic (water-loving) properties can be adjusted. Chronospheres are completely insoluble in water and other common solvents and can be incorporated into traditional liquid delivery vehicles (aerosol cans, sticks, pump sprays, etc.).

Many pharmaceutical active ingredients are sensitive to chemical and environmental exposure, resulting in degradation. Chronospheres are produced without the use of heat, water or solvents, thereby protecting ingredients during processing. In addition, no materials other than pure prepolymer and copolymerizable additives are used during Chronosphere™ production so there is minimal leaching except for the desired active material.

Confidential   **PolyMedica Industries, Inc.**

3.4 Markets

Chronosphere™ microparticulates are intended for use in the following areas:

Over-the-Counter (OTC) Pharmaceuticals

The total OTC pharmaceuticals market in 1989 was estimated at $9.7 billion by Nielson, an increase of 6.2% from the previous year. External remedies were at $1.6 billion, up 11.1%. Within this category, the following product formats could utilize Chronospheres to achieve superior efficacy via sustained/controlled release:

OVER-THE-COUNTER PHARMACEUTICAL MARKETS: 1988

	Estimated market size ($ millions)
Antiperspirants/deodorants	$ 600
Acne	250
External analgesics	240
External antiseptics	189
Hydrocortisones	90
Burn remedies	35
Poison ivy	32
Jock itch	9
	$1,445

Skin Care

The total skin care category reached $2 billion in 1989, an increase of 8.5% over 1988. Chronosphere™ products may be able to address the following product segments:
Suntan lotions
Moisturizers
Hand and body lotions
Face creams
Lip balms
Astringents

Fragrances

The fragrance market is estimated at $4 billion annually. Chronosphere™ products may be able to address the following areas of interest:

Dry scented products
Fragrance gels
Perfumes.

Industrial Applications

Chronosphere™ microparticulates offer sustained time-release and unique delivery characteristics. Their manufacturing cost efficiency would be attractive to the following industrial market segments:

Pesticides/ insecticides
Catalysts
Oxidizers
Pigments/colorants
Flame retardants
Water treatment
Biocides.

With such a wide variety of applications for this technology, PolyMedica is discussing the applications of its products in personal care markets with a well-positioned corporate partner.

4.0 CHRONOTHANE™ PHOTOPOLYMERS

4.1 Sustained-Delivery Technology

The past few years have brought a virtual revolution in delivery systems used by the pharmaceutical industry. Many of the common ointments and capsules have been replaced by transdermal patches. These deliver an active ingredient through the skin of the patient directly into the body. Future developments promise even more dramatic changes in drug delivery.

The impetus for delivery changes is an effort to eliminate the limitations of existing therapies. Sustained-delivery products improve drug performance by:

Increasing duration of action
Improving patient compliance
Eliminating side effects.

Secondly, the Price Competition and Patent Term Restoration Act of 1984 has substantially changed the competitive character of the pharmaceutical industry. Under this law, a new dosage that enhances the therapeutic efficacy of an existing drug may receive a separate patent in its own right.

This, in essence, would grant an additional period of patent exclusivity to a drug company that took one of its proprietary drugs, whose patent was expiring or had expired, and put it in a new delivery vehicle. Thus, a new generation of controlled/sustained delivery technologies has begun. Ciba-Geigy was one of the first companies to introduce transdermal patches into the market.

The magnitude of the market has attracted companies such as Alza, Elan, Hereon, Ciba-Geigy, Schering-Plough, and others. In each case, the companies have relied on proprietary polymer based systems which formed the basis for their core delivery technology.

4.2 Chronothane™ Sustained Delivery Products

In October 1988, PolyMedica applied for a patent on a novel phatopolymer to commercially exploit sustained-release products based on proprietary polymer technology. The products can be constructed to deliver in a topical, transdermal, transmucosal, or intraoral system. Topical delivery involves local delivery of medications within a specific site of the skin. In the transdermal mode, medications are delivered across intact skin. Transmucosal refers to the delivery of drugs through the lining of the cheek, and intraoral refers to delivery of medications within the mouth.

The medications envisioned by PolyMedica for its unique drug delivery systems are: (a) OTC preparations and (b) prescription drugs (Such as antibiotics, antimicrobials, and systemic medications which require continuous dosing, such as AZT). These incorporated active ingredients are placed into a polymeric-controlled release layer. They are delivered directly to the skin in a preprogrammed manner. In other words, the actives will be trapped in a polyurethane matrix designed to release them out at a specific dosage over a predetermined time period. PolyMedica has developed the fundamentals for drug delivery systems for the human health care market using the combined disciplines of polymer chemistry, pharmaceutics, and web handling engineering.

The Company's strategy for commercializing its unique technological ability in controlled and sustained release products is composed of two elements:

1. Expanding PolyMedica's proprietary polymer technology and maintaining ownership of the technology through patent and corporate secrecy protection. This means that in early stage product development, the Company may fund research and development expenses in its own laboratories.

2. Establishing long-term strategic relationships with leading companies for the development, marketing and sale of their own products which could benefit from patent protection if incorporated in a new delivery system.

4.3 Markets

According to Business Communications Company, a Stamford, Connecticut-based market research consulting firm, the estimated annual U.S. sales of new drug delivery systems from manufacturer to distributor grew from $720 million in 1983 to $1.8 billion in 1988. Drug delivery system sales are expected to soar to $3.5 billion by 1993. At the present time, Alza, Key Pharmaceuticals, and several other publicly-traded companies have the major share of the market for new drug delivery systems. Jim McCamant of the California-based *Medical Technology Stock Letter* says, "The potential market is still very large—large enough to make any company with a unique and efficient drug delivery production system prosperous."

The Company's product development efforts for Chronothane™ have emphasized two areas of human health care: transdermal delivery and topical (local) drug delivery.

Confidential   **PolyMedica Industries, Inc.**

4.4 The PolyMedica Competitive Advantage

PolyMedica's polymer technology and its applications in pharmaceuticals are at the forefront of controlled-release research. The Company has developed a unique polymer system, trade named Chronothane™, which forms the technology core for initial products.

Several distinguishing features set PolyMedica apart from its competition:

1. The Company is the only group known to produce hydrophobic and hydrophilic photopolymerizable polyurethane elastomers. These polymers, when used as the matrix layer, are capable of delivering hydrophobic and hydrophilic drugs at controllable rates.
2. Chronothane™ photopolymer films are produced at room temperature, in the absence of water and solvents. Heat, water and solvents are known to degrade many drugs. The Chronothane™ matrixes are produced in the absence of heat or water, thus ensuring full pharmacological activity and stability of the drug imbedded in the matrix layer.

4.5 Present Competitive Products

Below is a summary of the most active companies in the field of sustained/controlled delivery systems, with existing and/or envisioned new products.

COMPANY	PRODUCT(S)	MARKETING PARTNER	INTRODUCTION
Advanced Polymer Systems Redwood City, CA	Microsponges	Sterling Drug, Scott Paper	1990s
Alza Palo Alto, CA	Scopolamine, nitroglycerine, estradiol	Ciba-Geigy	Early 1980s
Bend Research Bend, OR	Oral and implant polymer tech.	Upjohn	Undetermined

COMPANY	PRODUCT(S)	MARKETING PARTNER	INTRODUCTION
Daltex Medical W. Orange, NJ	Drug coatings medical devices	C.R. Bard, Block Drug	Early 1990s
Elan Monksland, Ireland	Nicotine	None	Late 1980s
Eurand America Valhalla, OH	Polymer-based capsules	A.H. Robbins	Late 1980s
Hereon Labs S. Plainsfield, NJ	Nitroglycerine	Bolar Pharm.	Mid 1980s
Key Pharm. Miami, FLA	Nitroglycerine	None	Early 1980s
Nova Pharm. Baltimore, MD	Implant	None	Early 1990s

4.6 Benefits to the Pharmaceutical Industry

One of the most pressing problems facing the pharmaceutical industry is that in the past few years only a very limited number of new drug products have been approved for marketing by the Food and Drug Administration ("FDA"). Pharmaceutical companies in the U.S. spent $1 billion for R&D in 1975 alone, but produced only seven new drug entities. In contrast, in the mid 1950s, about forty new drug entities were introduced yearly by U.S. pharmaceutical companies with total annual R&D expenditures of less than $100 million. As patents on some 150 pharmaceuticals expire in the coming few years, suppliers to the $20 billion dollar a year prescription drug market are facing competition from makers of generic compounds. To prevent a potential revenue loss, industry leaders are evaluating the feasibility of incorporating new drug delivery techniques into many of their largest selling products. If they are successful, they will replace obsolete drug patents with new patents that protect the methods by which their existing drugs are administered.

PolyMedica has developed a family of novel photopolymerizable elastomers which can be used advantageously in a variety of controlled-release devices.

This family of proprietary photopolymers can incorporate prescription and over-the-counter drugs without resorting to heat, solvents, or water. This is a crucial consideration with many drugs which are easily degraded by heat, solvents, or water. In addition, these polyurethane elastomers are manufactured in a continuous web process, insuring high reproducibility and low manufacturing costs. These key economic features reflect a state-of-the-art advancement over current methods.

4.7 Development and Approval Process

Some of the Company's delivery system products will need to be approved by the FDA prior to being sold in the U.S. There may exist comparable regulatory agencies in certain other countries when selling outside the U.S.

The Company's initial controlled delivery products are likely to be classified as a "device" rather than a "drug". Generally, the FDA approval process to determine the safety and efficacy of a "drug" takes many years and requires an IND (Investigational New Drug). This is a very expensive undertaking and PolyMedica's corporate partners would fund an IND. If the Company develops a "device", the FDA approval process is truncated, and there may be circumstances in which PolyMedica would fund the cost.

4.8 Joint Venture Partner Strategy

The technology of PolyMedica has applicability to the development of controlled release products in many therapeutic categories. The present strategy is to concurrently pursue a number of joint development contracts and product design agreements with major pharmaceutical companies. In recent years, such collaborative agreements have been commonly used in the pharmaceutical industry to implement programs where smaller companies want to allow major drug companies access to their technology.

The Company expects specific terms to vary from partner to partner, but Poly Medica will seek agreements which will provide it with: (1) immediate revenues under development contracts, (2) manufacturing rights for products or polyurethane material components developed by the Company, and (3) royalties based on product sales. In turn, the joint venture partners will be granted exclusive rights in a particular geography for the sale and distribution of jointly-developed products.

POLYURETHANES

A polymer is a large molecule constructed from many smaller structural units called monomers, which are covalently bonded together. Polyurethanes are among the most important classes of polymers because of their unique physical and chemical properties. In addition to a significant number of urethane groups, polyurethanes may contain aliphatic and aromatic hydrocarbons, ester, ether, amide, and urea groups.

Polyurethane was originally developed by Professor Otto Bayer in 1937, who was trying to develop fibers to compete with nylon. In the following year, the DuPont Company developed a polyester type polyurethane based on diisocynate chemistry and it became the basis of most products made from polyurethanes for the next quarter of a century. In 1962, Johannes Mirkovitch, Ph.D. first used the polyester polyurethane for making artificial organs. Unfortunately, this polyester material was attacked by the body's enzyme systems and hydrolyzed.

The use of polymers in medical applications and devices continued to increase over the years as the search went on for a more stable polyurethane for implantable devices. In 1967, John W. Boretos, Ph.D. and William Pierce, M.D., Ph.D. introduced Biomer[R], a segmented polyether-type polyurethane, for biomedical applications such as artificial hearts. Biomer[R] became the yardstick by which all other polyurethanes were measured because it had the best flexure endurance characteristic of any polyurethane. In 1971, Emery Nyilas, Ph.D. developed a material called Avcothane[R] which advanced polyurethanes through the addition of silicone and made the polyurethane even more blood compatible.

In 1977, the Upjohn Company introduced a thermoplastic polyurethane, trade named Pellethane[R], as an alternative biomedical material. Because Pellethane[R] is a thermoplastic, it could be processed more cheaply and made more dimensionally precise. Concern over the implantation of polyurethanes began to increase in the late 1970s because of harmful substances which might be released during biodegradation. Then, in 1978, Thermedics, Inc. introduced an aliphatic polyurethane called Tecoflex[R]. Because of its chemical structure, Tecoflex[R] could not form carcinogenic amines if it degraded in the body. However, Tecoflex [R] is not suitable for long-term implants.

Confidential   **PolyMedica Industries, Inc.**

The process by which Biomer[R], Avcothane[R], Pellethane[R], and Tecoflex[R] have been shown to undergo *in vivo* degradation phenomena is known as environmental stress cracking (ESC), or microcracking. Such microcracking can lead to catastrophic failures in surgically implanted devices because the material weakens and eventually fractures. This ends all functionality of the device.

PolyMedica scientists have developed a proprietary biostable polyurethane elastomer which will not microcrack when used *in vivo*. This is an important advance which combines the best attributes of previous polyurethanes and significantly advances polymer chemistry. The PolyMedica polyurethane has three important attributes. Firstly, it is a biostable polymer which will not microcrack. Secondly, it is aliphatic and cannot form carcinogenic amines. Thirdly, it is thermoplastic, so it can be processed by all conventional plastic manufacturing techniques.

TABLE I

<u>POLYMERS ACCEPTED FOR USE IN MEDICAL APPLICATIONS</u>

<u>POLYMER</u>	<u>APPLICATION</u>
Acrylics	Hemostatic agents, bone replacement, corneas, facial prostheses
Cellulose acetate	Nerve regeneration, packaging material
Fluorocarbons	Blood vessels, reconstructive surgery
Polyamides	Syringes, blood transfusion sets, vascular implants
Polycarbonates	Syringes, containers
Polyethylene	Tubing, syringes, heart valves, knee joints
Polypropylene	Syringes, implants, bones and joints
Polyurethanes	Plastic surgery, biodegradable sutures, artificial hearts, catheters
Polyvinyl chloride	Surgical tubing, blood collection and administration, blood vessel sets
Silicones	Tubing, catheters, lubricants, tissue substitutes, heart and related components

TABLE II

CURRENT DEVICES AND THEIR COMPOSITION

DEVICE	FUNCTION	BIOMATERIAL IN USE
Artificial vitreous humor	Fill the vitreous cavity of the eye	Silicone, Teflon sponge, polyglycerylmethacrylate (PGMA)
Corneal prostheses	Provide an optical pathway to the retina	Polymethylmethacrylate (PMMA), hydrogels
Intraocular lens	Correct problems caused by cataracts	PMMA (lens), nylon, polypropylene, Pt, Ti, Au loops
Artificial tear duct	Correct chronic blockage	PMMA
Artificial eustachian tube	Provide clear ventilation passage	Silicone rubber, Teflon
Nerve tubulation	Align severed nerves	Silicone membrane, porous surgical metals
Middle ear prostheses	Replace diseased bones of the middle ear	PMMA, metallic wire, Proplast (PTFE + carbon fiber), Bioglass
Percutaneous leads	Conduct power to electrical sensory devices	Nylon or Dacron velour, PMMA
Auditory prostheses, visual prostheses	Restoration of hearing and vision	Pt and Pt-Ir wires and electrodes, Ta-\TOa$_2$O$_5$ electrodes, stainless steel, Elgiloy wires, silicone rubber, PMMA
Electrical analgesia	Eliminate chronic pain	Same
Electrical control of epileptic seizure	Conduct electrical signals to brain	Same
Electrophrenic stimulation	Control breathing electrically	Same
Bladder control	Stimulate bladder release	Same

TABLE III

SYNTHETIC POLYMERS

POLYMER	APPLICATION
Polyvinyl chloride	Blood bags, bubble oxygenators, oxygenator and dialysis tubing, catheters
Polyethylene (high molecular weight)	Orthopedic, acetabular cup
Polyethylene terephthalate	Woven vascular prostheses, heart valve suture rings
Polyacrylonitrile (PAN)	Dialysis membrane
Polytetrafluoroethylene (TeflonR TFE)	Orthopedic, coating stem prostheses, coating tips, neurosurgery, aneurysm clips
Polytetrafluoroethylene-hexafluoropropylene copolymer (Teflon^R FEP)	Orthopedics
Polycyanoacrylates	Tissue adhesives
Polymethylmethacrylate	Neurosurgery, cranioplasty, orthopedics, grout for artificial joints, opthamology lenses, contact lenses, dental
Polyhydroxyethylmethacrylate	Antithrombogenic surfaces, tubing, plastic surgery, drug release, contact lenses
Polydihydroxypropyl methacrylate	Antithrombogenic surfaces, tubing, plastic surgery, drug release, contact lenses
Resin: diglycidyl ether of bisphenol A (DGEBA)	Packaging pacemakers, central restorations

TABLE III (continued)

SYNTHETIC POLYMERS

Polyurethanes (segmented)	Catheters, tubing, artificial heart ventricles, pumps
Polyglycolic acid	Resorbable sutures
Silicones	Heart valves, tubing, plastic surgery, oxygenators
Silicone-polycarbonate copolymer	Oxygenator membranes
Copolymer silicone-polyurethane (Avcothane[R])	Left ventricular assist devices, intra-aortic balloon pumps
Polyphenoxy	Oxygenator membranes
Polyamide (Nylon 6)	Sutures
Polyformaldehyde (Delrin[R])	Heart valve occluders, general structural polymers
Epichlorohydrin rubber (Hydrin[R])	Flexible molds, impression material, dental applications
Polycarbonate (Lexan[R])	General use oxygenator valve occluders, cranioplasty
Cellulose	Dialysis membranes
Cellulose acetate	Dialysis membranes
Polyalkylsulfone	Membrane oxygenators
Polypropylene	Syringes, catheters, oxygenators, dialysis, heart valves, occluders, sutures
Polyacrylic acid	Dental cement
Gelatin/resorcinol/formaldehyde	Tissue adhesive
Polysulfide rubber (mercaptan)	Dental impressions

PolyMedica Industries, Inc.

PATENTS

PolyMedica has retained Morgan & Finnegan, a New York City-based law firm, as its patent counsel. At the present time, the Company has four issued patents, three patents pending, and numerous foreign analogues.

ISSUED PATENTS

4,704,130 <u>Biocompatible Microporous Polymeric Materials and Methods of Making Same</u>. November 3, 1987.

> This is a process patent describing the formation of the microporous membrane. A process patent is granted on a new, useful, and unobvious technical process.

> The process describes the formation of microporolus flat sheets, or hollow fibers, by casting a desired thickness of solvent-based "Mitrathane™" polyurethane, followed by precipitation in water. The solvent is subsequently extracted and the article is dried by heat.

4,813,966 <u>Biocompatible Microporous Polymeric Materials and Methods of Making Same</u>. March 21, 1989.

> This is a divisional application of 4,704,130 above.

> This patent covers the shape and size of porous structures and dimensions obtainable with the casting/precipitation/extraction drying method.

> The sheets covered in the patent have: (1) a void volume of 50–80%,(2) a surface layer pore size of 0.1 to 28 microns, and (3) a contiguous layer of finger-like voids of 100 to 500 microns, extending normal to the surface skin layer.

4,849,458 <u>Segmented Polyether Polyurethane</u>. July 18, 1989.

> This is a composition-of-matter patent. Composition of matter patents are obtained on new, useful and unobvious chemical compounds and formulae.

> The novel polyurethane is a hydrophilic polymer, containing pseudocrosslinks, which surprisingly becomes stronger when wetted with water.

Confidential   **PolyMedica Industries, Inc.**

This is important, since wound dressings need to absorb large quantities of exudate and retention of physical properties is necessary to prevent rupture during removal of a wound dressing.

4,906,240 <u>Adhesive-Faced Porous Absorbent Sheet and Method of Making Same</u>. March 6, 1990.

This is a product, or design, patent. Product patents are obtained on new, useful, and unobvious product constructions.

The patent describes two structures: (1) a bilaminate structure composed of an absorbent porous sheet and a porous adhesive, and (2) a trilaminate structure composed of a liquid-impermeable film, an absorbent porous sheet, and a porous adhesive.

Further, the patent describes an adhesive layer with an aperture opening of one to fifteen percent, where fragments of the adhesive are retained within the pores of the macroporous sheet without blocking fluid channels.

PATENTS PENDING

<u>Adhesive-Faced Porous Absorbent Sheet and Method of Making Same</u>. Divisional Application of 4,906,240.

The application describes a process for manufacturing an adhesive-faced laminate, suitable as wound dressing, burn dressing, surgical drape, etc.

The patent discloses the process of making microporous adhesive laminated sheets by laminating, compressing and releasing an acrylic adhesive mass against a porous polyurethane membrane.

<u>Hydrophilic Polyurethane Elastomers</u>.

This is a composition-of-matter patent application. The application describes the synthesis of UV-curable elastomers. These elastomers are synthesized in the presence of a novel polymerizable photoinitiator, which becomes part of the molecular chain.

Because the photoinitiator becomes chemically bound into the elastomer, the finished polymer has essentially no extractables, an advantageous characteristic for use in drug delivery systems.

Confidential   **PolyMedica Industries, Inc.**

Further, the polymer can be formed in the absence of heat, since it cures upon exposure to UV radiation for a period of two minutes at room temperature.

<u>Biostable Polyurethane Products.</u>

A composition-of-matter application concerning polyurethane products possessing long-term biostability, in combination with a low modulus of elasticity, high ultimate strength and other desirable mechanical properties. These properties are crucial in medical devices such as pacemaker lead insulators, vascular grafts, mammary prostheses, and other medical products which are intended to be placed within the body for extended periods with no substantial degradation of the product.

<u>FOREIGN ANALOGUES</u>

It is the policy of PolyMedica to file patent analogues to its U.S. Patents in some or all of the following countries:

Australia
Canada
Europe:
   Austria
   Belgium
   France
   Germany
   Great Britain
   Greece
   Italy
   Netherlands
   Norway
   Spain
   Sweden
   Switzerland
Japan
Indonesia
Israel
New Zealand
Saudi Arabia
South Africa

Analogues will be filed in additional countries as appropriate markets develop.

A complete list of the Foreign Analogues of PolyMedica's patents has been omitted for reasons of strict confidentiality, but is available upon written request.

TRADEMARKS

Patent/Country	Name	Date Issued or Filed
U.S. Registered Trademark No. 1,584,191	Mitrathane™ polymer products	Issued on February 27, 1990
Canadian Trademark application No. 638,491		Published in Canadian Trademarks Journal, April 25, 1990
U.S. Trademark application No. 73/57816	Mitraflex™ Wound Dressings	Filed on October 17, 1988 (refiled)
Canadian Trademark application No. 637,121		Published in Canadian Trademarks Journal, March 14, 1990
Great Britain Trademark application No. 1,422,934		Filed on April 23, 1990
Spanish Trademark application No. 1,568,116		Filed on May 17, 1990

CAPITAL REQUIREMENTS

1. MASSACHUSETTS AND COLORADO FACILITIES

The Company is in the process of fitting out a 9,000 square foot manufacturing and laboratory facility in Woburn, Massachusetts and is upgrading its manufacturing plant and laboratories in Wheat Ridge, Colorado. The Company expects to have capital needs in the specific areas outlined below as well as the need for general working capital:

EQUIPMENT FOR CONTINUOUS PRODUCT MANUFACTURING

1.1 Wound Dressing

60" wide casting machine	$500,000
Form, fill and seal packager	60,000
Rewinder/slitter/inspection station	50,000
Rotary die cutter	50,000
	$660,000

At present, the Company produces Mitraflex™ in a 30" wide casting machine running at 2.5 ft./min. A 60" wide casting machine, run at 3.0 ft./min., will effectively double the Company's production capacity, thereby reducing manufacturing costs by more than half.

In addition, a wider membrane will allow wider casting/laminating and reduce conversion costs by half.

A form, fill and seal packager, in conjunction with a rotary die cutter, will quintuple our packaging capacity. Packaging is our current production bottleneck.

The rewinder is necessary to inspect rolls of converted Mitraflex™, prior to die cutting. It will also be used to slit the converted rolls into strips in preparation for rotary die cutting.

Confidential   **PolyMedica Industries, Inc.**

1.2 Biostable Polymer

Meter/mix and pouring station	$ 55,000
Myers high intensity emulsifier	9,000
Heating and filtering station	6,500
Blue M curing ovens (2)	11,000
Cumberland granulator	11,250
Killion pelletizing line	42,000
Hot house (Drum heater)	3,000
Polar winder	36,000
Extruder/calender/embossing line	67,000
	$240,750

The biostable polyurethane is synthesized from the starting monomers. The drums of monomers need to be heated and then: (a) emulsified, (b) filtered, (c) mixed under precise conditions, and (d) heat cured into cakes. The cakes are then granulated and pelletized.

The pellets are subsequently extruded into embossed sheets; or conversely, the pellets are extruded at an external source into yarns, which then are wound into fabrics by a polar winding machine.

1.3 Chronospheres

EPE UV-curing system	$ 90,000
Particulate conversion equipment	67,000
Granulator	11,250
Cowles dispersator line	30,000
Jacketed chemical reactor	60,000
Steam generator	7,500
	$265,750

Confidential   **PolyMedica Industries, Inc.**

Chronospheres are prepared from special photopolymers. An oligomer is synthesized in the steam-heat jacketed chemical reactor. PolyMedica requires a Cowles dispersator to incorporate the active(s) into the oligomer; the oligomer and actives are then cured by the UV curing system.

1.4 Fit Up of Facility

Ventilation, computers and packaging equipment	$ 50,000
Utilities	55,000
Clean room	28,500
Furniture and fixtures	100,000
	$233,500

Improvements in the manufacturing process at Wheat Ridge, Colorado for Mitraflex™ are necessary to reduce costs and increase efficiency. The Company estimates it will have to spend approximately $50,000 to upgrade ventilation equipment, change the physical flow of manufacturing, introduce new computer systems, and purchase new packaging equipment.

In Woburn, Massachusetts, the Company will need to install electrical outlets, running water and lights in the laboratories and production facilities.

A clean room is necessary to aseptically package devices under low bioburden conditions.

PolyMedica will have to purchase a telephone system, desks, chairs, filing cabinets, lights and the normal fit up for a facility.

Total for the Massachusetts and Colorado facilities $1,400,000

2. RETIREMENT OF ADDITIONAL WARRANT

The agreement dated June 15, 1990, under which the Company purchased the assets of Matrix Membranes, Inc., contained a warrant to purchase common stock which was granted to Technology Funding Partners III ("TFP"). Under the terms of the warrant, TFP has the right to purchase $1.4 million of stock at a price per share equal to 50% of the offering price of a future round of financing. This right was granted to TFP at the time of the asset purchase as part of the consideration paid. The Company will retire this warrant in a future round of financing.

The net effect of the retirement of this warrant is that PolyMedica will receive $700,000 of proceeds for the issuance of $1.4 million of Common Stock, which will be part of an offering. For example, the maximum cash proceeds to the Company for a $3.5 million round of financing would be $2.8 million if the warrant was exercised and retired.

3. SALES AND MARKETING

PolyMedica is establishing distribution of its Mitraflex™ and Chronothane™ products to numerous markets, namely hospitals, nursing homes, and physicians; veterinarians and veterinary hospitals; universities and affiliated educational institutions; and OEM manufacturers in such areas as desiccants, biosensors, textiles, and cosmetics. These markets are being developed both in the U.S. and in countries within Europe and the Far East. The primary method of marketing/selling will be to utilize established distribution companies that stock the product on a local basis and then sell to the end user via their own sales force. In the U.S., there are numerous national, regional, and local distributors. Through relationships with a variety of companies, all U.S. markets will be covered. In the international markets, it is customary for one company to represent an entire region and/or country. These companies establish an exclusive distribution arrangement for a period of time based upon minimum performance requirements.

In order to support these marketing efforts, it is necessary for PolyMedica to engage in telemarketing, direct mail and promotion programs and trade show exhibits, as well as hire a supporting sales force to work with the distributors in selling the products to end users. This sales force primarily serves as specialists to answer technical issues as well as assist in the closing process. Thus, the Company estimates that approximately $300,000 of the proceeds of a future offering will be used to implement various marketing activities.

4. PATENT PROTECTION

Given its unique technology, PolyMedica will apply for patents in the numerous markets which exist for its technology. (See "The Company and Its Products".) The Company will decide on foreign patent filings on an individual country and product basis.

In addition, the Company will actively defend its patents against infringement. PolyMedica expects to allocate approximately $150,000 of the proceeds of a future offering for patent legal and filing expenses.

5. SUMMARY OF CAPITAL NEEDS

The Company expects to have the following cash needs in the coming twelve months. An allocation summary of these cash proceeds is as follows:

Massachusetts and Colorado facilities	$1,400,000
Sales and marketing	300,000
Patent protection	150,000
Working capital	950,000
	$2,800,000

OPERATING PLAN

PRO FORMA SALES

OVERVIEW

PolyMedica is projecting a rapid increase in sales during the next five years due to the unique market position of its products (see "The Company and Its Products"). The Company believes that sales may grow from $5.3 million for the fiscal year ending ("FYE") March 31, 1992 to $40.6 million for FYE 1996. Current sales are derived from introductory sales of Mitraflex™ wound dressings and specific product research. Future sales growth includes derivative products of Mitraflex™ wound dressings to expanded U.S. markets as well as new sales of Mitraflex™ in Europe and the Pacific Rim. Additional targeted segments will be Mitraflex™ for veterinary markets, introduction of industrial and medical textile products, desiccant films, implants and expanders, transdermal systems, and cosmetics and personal care products.

MITRAFLEX™ AND EXTENSIONS IN THE U.S. MARKET

Sales of Mitraflex™ wound dressings for FYE 1992 to the Calgon Vestal Division of Merck are expected to total $1.8 million. These sales are expected to grow to $9.1 million by FYE 1996. In addition, the Company may begin generating $125,000 of sales of Mitraflex™ spin-off products in U.S. markets beginning in the quarter ending June 30, 1991, and should grow rapidly to an annual rate of $8.6 million during FYE 1996.

EUROPEAN AND PACIFIC MARKETS

During the first quarter of FYE 1992, PolyMedica expects to commit the resources to market Mitraflex™ products in Europe and the Pacific Rim. The Company anticipates this effort to result in a rapid growth curve to yield annual sales of $4 million in Europe and $2.5 million in the Pacific Rim by FYE 1996.

VETERINARY WOUND MARKETS

PolyMedica's entrance into the veterinary market should commence in the first quarter of FYE 1992 with $50,000 in sales. The Company expects this market to generate sales of $2.7 million annually by FYE 1996.

ADDITIONAL PRODUCT SALES

In addition to its Mitraflex™ products, PolyMedica's introduction of medical and industrial textile products is estimated to generate sales of $50,000 for the quarter ended September 30, 1991 and grow to $2.4 million by FYE 1996. Desiccant film sales, which are products made of the existing outer film, should begin in the first quarter of FYE 1992, and reach $245,000 during FYE 1992. They are expected to grow to $1.2 million annually by FYE 1996.

Sales of implants and expanders, which are based on the Company's Chronoflex™ materials, should equal $50,000 for the quarter ended December 31, 1991 and grow to $1.8 million annually by FYE 1996. Transdermal systems, a result of PolyMedica's Chronothane™ technology, may be sold at a $50,000 quarterly rate beginning in the quarter ended June 30, 1992 and could grow to $1.8 million by FYE 1996.

Finally, cosmetic and personal care products emanating from PolyMedica's Chronosphere™ microparticulate materials are expected to ramp up from a quarterly sales volume of $90,000 in the quarter ended September 30, 1992, to an annual volume of $2.1 million by FYE 1996. Management believes this is a very conservative projection.

PRODUCT RESEARCH

The Company will be conducting ongoing product research and development with partners. In each case, the Company will endeavor to maintain the right to manufacture some portion from critical components through the finished product. Product research income is expected to grow from an annual rate of $800,000 in FYE 1992 to $2.0 million by FYE 1996.

ROYALTIES AND LICENSE FEES

Royalty revenues and license fees earned from certain of the above product sales are anticipated to equal $450,000 in FYE 1992, and are projected to increase to $2.5 million by FYE 1996.

Confidential   **PolyMedica Industries, Inc.**

PRO FORMA STATEMENTS OF OPERATIONS

COMPANY PROFITABILITY

PolyMedica expects to show its first quarterly profit before taxes of $118,000 during the quarter ended March 31, 1992 and record an overall loss before taxes of $138,100 for FYE March 31, 1992. These results are based on sales of $1.8 million and $5.3 million, respectively.

Thereafter, the Company anticipates recording steadily increasing pretax profits, which are projected to be $1.2 million, $2.3 million, $6.0 million and $10.8 million for FYE 1993–1996, respectively. On a percentage of sales basis, this performance ranges from 11.1% to 26.5% for the above fiscal years.

Under the current plan, net operating loss carryforwards will defer income tax payments until quarter ended March 31, 1993.

GROSS MARGINS

As its sales grow, PolyMedica's gross margin will increase as manufacturing efficiencies are realized across a large spectrum of the Company's product lines. This fact is especially true with Mitraflex™ extensions, as significant increases in sales volume for numerous versions of Mitraflex™ products may be realized while incurring an incrementally smaller increase in manufacturing costs.

For FYE 1992 through FYE 1996, the Company's gross margins should increase from 55.9% to 68.0% of sales. Gross margin dollars that equate to these percentages are expected to grow from $2.9 million in FYE 1992 to $27.6 million by FYE 1996.

OTHER OPERATING EXPENSES

Other operating expenses, which are defined as research and development, marketing and sales, and general and administrative, represent a steadily declining percentage of sales as the Company's sales volume increases, margins grow, and other efficiencies are realized.

Research and development expenses are anticipated to equal $698,000 in FYE 1992 and grow to $4.3 million by FYE 1996. This increase reflects PolyMedica's commitment to remaining on the cutting edge in its field. As a percentage of sales, however, the trend is positive, as total research and development expenses may decrease from 13.3% in FYE 1992 to 10.6% by FYE 1996.

Marketing and sales expenses as a percentage of sales may remain relatively constant from FYE 1992 to FYE 1996, ranging from 18.6% to 16.6%. On a dollar basis, these amounts are expected to grow from $978,800 to $6.7 million for the above fiscal years. These increases could support the Company's aggressive marketing plans for: (1) Mitraflex™ extensions and entry into European and Pacific Rim markets and (2) the commercial realization of PolyMedica's Chronosphere™ polymer microparticulate and Chronothane™ biostable polyurethane technologies.

General and administrative expenses, as a percentage of sales, may drop substantially from FYE 1992 to FYE 1996. Projected percentages are expected to be 30.1% in FYE 1992, with a steady reduction to 14.9% by FYE 1996. These percentages equate to $1.6 million in FYE 1992, increasing to $6 million by FYE 1996.

PRO FORMA BALANCE SHEETS

INCREASE IN ASSETS

The Company is projecting a strong growth in its balance sheet, with total assets pro-jected to be $6.5 million at March 31, 1992 and increasing to $21.4 million by March 31, 1996. This positive trend is based on a significant sales increase, dramatically improved gross mar-gins and a disciplined approach to monitoring other operating expenses, specifically research and development, marketing and sales, and general and administrative.

INCREASE IN NET WORTH

Coincident with the rise in assets is the increase in PolyMedica's net worth. Shareholders' equity is estimated to equal $5.2 million as of March 31, 1992 and rise to $17.7 million as of March 31, 1996. These improvements in Company net worth will be accomplished without any external financing beyond a future round of financing.

COMPONENTS OF ASSETS

Current assets as a component of total assets, and therefore a measure of the Company's liquidity, are expected to increase from 53.9% as of March 31, 1992 to 79.5% of total assets as of March 31, 1996. This change reflects substantial increases in cash and work in process inventory. Gross property, plant and equipment, while decreasing as a percentage of total assets, should show an increase of $4 million from March 31, 1992 to March 31, 1996. The accumulation of depreciation and amortization expense reduces this increase to $844,000. Other assets remain in the 4%–8% range of total assets.

COMPONENTS OF LIABILITIES AND SHAREHOLDERS' EQUITY

Total liabilities as a percentage of liabilities and shareholders' equity decrease slightly from 21.2% as of March 31, 1992 to 17.4% as of March 31, 1996. These liabilities exclude external debt other than normal trade payables and accrued liabilities and reflect ongoing paydown of the Company's existing $400,000 promissory note. Therefore, as the Company grows, it should be able to fund increases in inventory and fixed assets without incurring additional leverage.

Under its operating plan, PolyMedica's net worth should remain positive from inception. After a future offering, the Company's additional paid-in capital should stay unchanged at $6.3 million. The accumulated deficit turns positive as retained earnings during early FYE 1994 and increases to $11.4 million by March 31, 1996.

PRO FORMA CASH FLOWS

PolyMedica's cash balances will be approximately $6.8 million as of March 31, 1996 based on: (1) cash generated from operations, (2) prior net financing totalling $2.4 million, and (3) possible generation of $2.8 million in cash from a future round of financing. There are no additional sources of financing proposed under the Company's plan.

CASH FLOW POSITIVE

The Company becomes cash flow positive from operations beginning in the quarter ended June 30, 1992, with a $470,000 increase in cash during the quarter. Positive cash flow should equal $1.3 million, $740,000, and $3.2 million for FYE 1994, 1995 and 1996, respectively. As the Company bears the full burden of income taxes beginning in FYE 1995, the rate of its annual positive cash flow decreases by $535,000 compared with FYE 1994. This trend, however, may dramatically reverse itself FYE 1996.

RECEIPTS

Receipts from accounts receivable, estimated to be $4.4 million in FYE 1992, are expected to rise to $38.4 million annually by FYE 1996.

DISBURSEMENTS

In addition to the normal disbursements for direct and operating expenses, the plan shows significant growth in disbursements for (1) inventory, (2) manufacturing and laboratory equipment, and (3) patent and trademark fees. These disbursements total $1.6 million in FYE 1992 and increase to $2.1 million by FYE 1996.

Confidential   **PolyMedica Industries, Inc.**

APPENDIX A

AUDITED FINANCIAL STATEMENTS FOR

THE PERIOD ENDED MARCH 31, 1990

These will be available by written request.

APPENDIX B

Detailed Financial Projections:

FYE March 31, 1992–1996

PolyMedica Industries, Inc.
Pro Forma Sales

PolyMedica Industries, Inc.

($000s)	Q/E 06/30/91	Q/E 09/30/91	Q/E 12/31/91	Q/E 03/31/92	Total FYE 1992	Q/E 06/30/92	Q/E 09/30/92	Q/E 12/31/92	Q/E 03/31/93	Total FYE 1993	FYE March 31, 1994	FYE March 31, 1995	FYE March 31, 1996
A. PRODUCT SALES:													
MITRAFLEX													
Calgon:													
4x4	$250	$280	$315	$360	$1,205	$400	$440	$480	$520	$1,840	$2,580	$3,600	$4,700
8x8	100	125	150	175	550	200	220	240	260	920	1,290	1,800	2,350
Other	–	15	25	45	85	65	80	95	110	350	700	1,300	2,000
Total Calgon	350	420	490	580	1,840	665	740	815	890	3,110	4,570	6,700	9,050
Mitraflex Extensions -US	125	140	190	230	685	275	325	375	425	1,400	2,800	4,900	8,600
Mitraflex and Extensions - International:													
Europe	100	115	130	145	490	160	180	200	220	760	1,500	2,800	4,000
Pacific Rim	25	25	45	60	155	75	90	115	140	420	800	1,500	2,500
Total International	125	140	175	205	645	235	270	315	360	1,180	2,300	4,300	6,500
Mitraflex -Veterinary	50	50	70	90	260	110	130	150	170	560	1,000	1,750	2,650
TOTAL MITRAFLEX	650	750	925	1,105	3,430	1,285	1,465	1,655	1,845	6,250	10,670	17,650	26,800
TEXTILES	–	50	75	90	215	125	150	175	225	675	1,200	1,800	2,400
DESICCANT FILM	30	50	75	90	245	105	120	130	140	495	750	1,000	1,200
IMPLANTS/EXPANDERS	–	–	50	75	125	90	110	130	150	480	750	1,200	1,800
TRANSDERMAL SYSTEMS	–	–	–	–	–	50	75	90	110	325	600	1,200	1,800
COSMETICS AND PERSONAL CARE	–	–	–	–	–	–	90	110	140	340	900	1,500	2,100
TOTAL PRODUCT SALES	680	850	1,125	1,360	4,015	1,655	2,010	2,290	2,610	8,565	14,870	24,350	36,100
B. PRODUCT RESEARCH	180	180	220	220	800	220	220	260	260	960	1,300	1,600	2,000
C. ROYALTIES	42	53	70	85	250	104	126	144	163	537	905	1,459	2,063
D. LICENSES	–	100	–	100	200	–	125	–	125	250	300	350	400
GRAND TOTAL	$902	$1,183	$1,415	$1,765	$5,265	$1,979	$2,481	$2,694	$3,158	$10,312	$17,375	$27,759	$40,563

Confidential

PolyMedica Industries, Inc.
Pro Forma Statements of Operations

($000s)	FYE March 31, 1992					FYE March 31, 1993					FYE March 31, 1994	FYE March 31, 1995	FYE March 31, 1996
	Q/E 06/30/91	Q/E 09/30/91	Q/E 12/31/91	Q/E 03/31/92	Total FYE 1992	Q/E 06/30/92	Q/E 09/30/92	Q/E 12/31/92	Q/E 03/31/93	Total FYE 1993			
Sales	$901.6	$1,183.3	$1,415.1	$1,764.8	$5,264.8	$1,978.5	$2,481.4	$2,693.6	$3,158.9	$10,312.4	$17,375.3	$27,758.8	$40,562.5
Cost of Sales	454.4	530.4	630.6	708.6	2,324.0	801.6	920.6	1,030.8	1,117.8	3,870.8	5,896.0	9,117.0	12,960.0
Gross Margin	447.2	652.9	784.5	1,056.2	2,940.8	1,176.9	1,560.8	1,662.8	2,041.1	6,441.6	11,479.3	18,641.8	27,602.5
Operating expenses:													
Research and development	140.4	157.7	187.8	211.9	697.8	257.9	292.2	328.4	370.7	1,249.2	2,058.4	3,057.4	4,316.9
Marketing and sales	183.3	214.6	274.6	306.3	978.8	390.2	422.6	494.2	536.6	1,843.6	3,451.0	5,169.0	6,740.0
General and administrative	339.1	385.9	412.2	448.9	1,586.1	500.3	541.5	605.1	663.3	2,310.2	3,739.0	4,669.0	6,046.0
Total operating expenses	662.8	758.2	874.6	967.1	3,262.7	1,148.4	1,256.3	1,427.7	1,570.6	5,403.0	9,248.4	12,895.4	17,102.9
Interest income	67.6	56.5	44.5	36.3	204.9	31.4	40.8	36.2	33.8	142.1	130.1	231.9	290.7
Interest expense	0.0	(7.0)	(7.0)	(7.0)	(21.0)	(7.0)	(7.0)	(7.0)	(7.0)	(28.0)	(28.0)	(28.0)	(28.0)
Profit (loss) before income taxes	67.6	49.5	37.5	29.3	183.9	24.4	33.8	29.2	26.8	114.1	102.1	203.9	262.7
Provision for income taxes	(148.0)	(55.9)	(52.5)	118.3	(138.1)	52.9	338.2	264.3	497.3	1,152.7	2,333.0	5,950.2	10,762.3
Net income (loss)	($148.0)	($55.9)	($52.5)	$118.3	($138.1)	$52.9	$338.2	$264.3	$479.2	$1,134.6	$1,392.8	$3,552.3	$6,425.1

Confidential

PolyMedica Industries, Inc.

PolyMedica Industries, Inc.
Pro Forma Balance Sheets

($000s)	Jun30 1991	Sep30 1991	Dec 31 1991	Mar 31 1992	Jun30 1992	Sep 30 1992	Dec 31 1992	Mar 31 1993	Mar 31 1994	Mar 31 1995	Mar 31 1996
ASSETS											
CURRENT ASSETS											
Cash and cash equivalents	$2,824.2	$2,226.3	$1,813.6	$1,568.3	$2,038.2	$1,808.8	$1,689.3	$1,626.8	$2,898.2	$3,633.7	$6,785.6
Accounts receivable	601.1	788.8	943.4	1,176.5	1,319.0	1,654.3	1,795.8	2,105.9	2,895.9	4,626.5	6,760.4
Prepaid expenses	50.0	50.0	65.0	75.0	100.0	120.0	140.0	160.0	180.0	200.0	220.0
Inventory:											
Raw materials	90.9	106.1	126.1	141.7	160.3	184.1	206.2	223.6	294.8	455.9	648.0
Work in process	295.4	344.8	409.9	460.6	521.0	598.4	670.0	726.6	958.1	1,481.5	2,106.0
Finished goods	68.2	79.6	94.6	106.3	120.2	138.1	154.6	167.7	221.1	341.9	486.0
Total inventory	454.4	530.4	630.6	708.6	801.6	920.6	1,030.8	1,117.8	1,474.0	2,279.3	3,240.0
TOTAL CURRENT ASSETS	3,929.6	3,595.6	3,452.6	3,528.4	4,258.8	4,503.6	4,655.8	5,010.5	7,448.1	10,739.4	17,006.0
PROPERTY AND EQUIPMENT											
Laboratory equipment	766.9	861.4	901.9	942.4	1,021.4	1,048.4	1,088.9	1,156.4	1,385.9	1,635.7	1,905.7
Manufacturing equipment	1,158.5	1,453.0	1,619.5	1,787.5	1,817.5	1,882.5	1,980.0	2,226.5	2,627.8	3,403.1	4,291.1
Office equipment, furniture and fixtures	173.5	191.0	198.5	206.0	216.0	221.0	228.5	241.0	283.5	329.8	379.8
Leasehold improvements	369.1	379.6	384.1	388.6	394.6	547.6	702.1	709.6	735.1	762.9	792.9
Less accumulated depreciation and amortization	(456.5)	(559.6)	(670.4)	(789.2)	(912.4)	(1,044.5)	(1,187.3)	(1,342.1)	(2,061.0)	(2,936.9)	(3,989.6)
TOTAL PROPERTY AND EQUIPMENT	2,011.5	2,325.4	2,433.6	2,535.3	2,537.1	2,655.0	2,812.2	2,991.4	2,971.3	3,194.4	3,379.7
OTHER ASSETS											
Organizational costs, net of accumulated amortization	27.0	24.0	21.0	18.0	15.0	12.0	9.0	6.0	0.0	0.0	0.0
Patents and Patent application costs	440.0	460.0	500.0	500.0	540.0	580.0	600.0	620.0	780.0	930.0	1,080.0
Less accumulated amortization	(26.5)	(29.0)	(31.5)	(33.9)	(36.8)	(40.0)	(43.8)	(48.2)	(60.6)	(65.6)	(71.1)
Net patents	413.5	431.0	468.5	466.1	503.2	540.0	556.2	571.8	719.4	864.4	1,008.9
TOTAL OTHER ASSETS	440.5	455.0	489.5	484.1	518.2	552.0	565.2	577.8	719.4	864.4	1,008.9
	$6,381.6	$6,376.0	$6,375.7	$6,547.9	$7,314.1	$7,710.6	$8,033.2	$8,579.8	$11,138.8	$14,798.3	$21,394.6

PolyMedica Industries, Inc.
Pro Forma Balance Sheets

($000s)	Jun30 1991	Sep 30 1991	Dec 31 1991	Mar 31 1992	Jun30 1992	Sep 30 1992	Dec 31 1992	Mar 31 1993	Mar 31 1994	Mar 31 1995	Mar 31 1996
LIABILITIES AND SHAREHOLDERS' EQUITY											
CURRENT LIABILITIES											
Accounts payable	$471.8	$508.8	$547.8	$588.3	$901.4	$946.3	$991.4	$1,045.3	$1,823.9	$1,897.9	$1,985.9
Accrued liabilities	357.5	387.5	417.5	447.5	864.5	894.5	924.5	954.5	1,409.0	1,509.0	1,659.0
TOTAL CURRENT LIABILITIES	829.3	896.3	965.3	1,035.8	1,765.9	1,840.8	1,915.9	1,999.8	3,232.9	3,406.9	3,644.9
Promissory note	400.0	383.3	366.6	349.9	333.2	316.5	299.8	283.1	216.3	149.5	82.7
TOTAL LIABILITIES	1,229.3	1,279.6	1,331.9	1,385.7	2,099.1	2,157.3	2,215.7	2,282.9	3,449.2	3,556.4	3,727.6
SHAREHOLDERS' EQUITY											
Series A redeemable convertible preferred stock, $.01 par value	0.9	0.9	0.9	0.9	0.9	0.9	0.9	0.9	0.9	0.9	0.9
Series B redeemable convertible preferred stock, $.01 par value	5.2	5.2	5.2	5.2	5.2	5.2	5.2	5.2	5.2	5.2	5.2
Series C redeemable convertible preferred stock, $.01 par value	2.4	2.4	2.4	2.4	2.4	2.4	2.4	2.4	2.4	2.4	2.4
Common stock, $.01 par value	3.8	3.8	3.8	3.8	3.8	3.8	3.8	3.8	3.8	3.8	3.8
Additional paid-in capital	6,289.8	6,289.8	6,289.8	6,289.8	6,289.8	6,289.8	6,289.8	6,289.8	6,289.8	6,289.8	6,289.8
Retained earnings (deficit)	(1,149.8)	(1,205.7)	(1,258.3)	(1,139.9)	(1,087.1)	(748.8)	(484.5)	(5.3)	1,387.5	4,939.8	11,364.9
TOTAL SHAREHOLDERS' EQUITY	5,152.3	5,096.4	5,043.8	5,162.2	5,215.0	5,553.3	5,817.6	6,296.8	7,689.6	11,241.9	17,667.0
	$6,381.6	$6,376.0	$6,375.7	$6,547.9	$7,314.1	$7,710.6	$8,033.2	$8,579.8	$11,138.8	$14,798.3	$21,394.6

Confidential

PolyMedica Industries, Inc.

PolyMedica Industries, Inc.
Pro Forma Cash Flows

($000s)	Q/E 06/30/91	Q/E 09/30/91	Q/E 12/31/91	Q/E 03/31/92	Q/E 06/30/92	Q/E 09/30/92	Q/E 12/31/92	Q/E 03/31/93	FYE March 31, 1994	FYE March 31, 1995	FYE March 31, 1996
Beginning cash balance	$3,380.3	$2,824.2	$2,226.3	$1,813.6	$1,568.3	$2,038.2	$1,808.8	$1,689.3	$1,626.8	$2,898.2	$3,633.7
Cash receipts:											
Accounts receivable	625.5	995.5	1,260.5	1,531.7	1,836.0	2,146.1	2,552.1	2,848.7	16,585.3	26,028.2	38,428.5
Interest income	67.6	56.5	44.5	36.3	31.4	40.8	36.2	33.8	130.1	231.9	290.7
Total cash receipts	693.1	1,052.0	1,305.1	1,567.9	1,867.4	2,186.9	2,588.3	2,882.5	16,715.4	26,260.0	38,719.2
Cash disbursements:											
Direct cost of sales and operating expenses less depreciation and amortization	661.1	1,180.1	1,388.9	1,551.5	1,015.8	2,038.6	2,308.9	2,526.3	13,022.8	21,131.5	29,004.6
Change in other asset accounts	588.2	446.0	305.2	238.0	358.0	354.0	375.2	377.0	1,386.2	1,900.3	2,130.8
Promissory note principal payments	0.0	16.7	16.7	16.7	16.7	16.7	16.7	16.7	66.8	66.8	66.8
Interest expense on promissory note	0.0	7.0	7.0	7.0	7.0	7.0	7.0	7.0	28.0	28.0	28.0
Income tax payments	0.0	0.0	0.0	0.0	0.0	0.0	0.0	18.0	940.2	2,397.9	4,337.2
Total cash disbursements	1,249.3	1,649.8	1,717.8	1,813.2	1,397.5	2,416.3	2,707.8	2,945.0	15,444.0	25,524.5	35,567.4
Ending cash balance	$2,824.2	$2,226.3	$1,813.6	$1,568.3	$2,038.2	$1,808.8	$1,689.3	$1,626.8	$2,898.2	$3,633.7	$6,785.6

Confidential

PolyMedica Industries, Inc.

Chapter 15

Harvesting

15.1 Introduction

Even before you start your entrepreneurial venture, you must strategize how you personally intend to exit. An exit strategy is the way in which a venture capitalist or business owner intends to monetize an investment. An exit strategy is also called a liquidity event.

Planning an exit strategy is one of the most misunderstood considerations of a business strategy, yet the exit strategy plays a crucial role in determining the strategic direction. Founders must have a definitive exit strategy before approaching investors, clearly reflected in their business plan. An entrepreneur's exit is both a career choice as well as a liquidation of a financial investment.[1]

This is called "**harvesting**." Harvesting is the term used to describe a major monetizing event from the business you created by taking the company public, selling it, merging with a larger company, outlicensing the technology, etc. Waking up one day as a multimillionaire and exiting your "baby" is a life-changing event. Are you prepared?

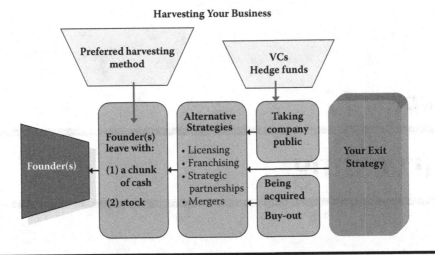

Figure 15.1   Start at the end—How do you personally intend to exit if successful?

15.2 Start at the End

"Money is not the only thing in life, but is way ahead of whatever is second."

Whether you are prepared, your investors want to know how they will benefit from their investment. Keep in mind that harvest refers to an event where the venture continues while the entrepreneur exits as both manager and major investor. Figure 15.1 summarizes some of the common strategies envisioned by founders prior to approaching potential investors.

15.2.1 Who Needs an Exit?

"A nickel ain't worth a dime anymore." —Yogi Berra

Entrepreneurs are dismayed when venture capitalists install their own CEO (and a new management team) just prior to going public. VCs generally consider entrepreneurs as technically qualified, but lacking in business skills. VCs argue that a public company requires seasoned business managers, not creative geniuses.

Founder/entrepreneurs exit successful ventures because of four major factors:

1. Facing new organizational performance standards
2. Desiring to cash in on the value created

3. Desiring to change lifestyles
4. Desiring to become a serial entrepreneur

15.3 Capitalization Principles

"I have seen that déjà vu before." —Yogi Berra

In business, **capitalization** means the following:

1. In accounting terms, it is where costs to acquire an asset are included in the price of the asset.
2. The sum of a corporation's stock, long-term debt, and retained earnings, also known as "invested capital."
3. A company's outstanding shares multiplied by its share price, better known as "market capitalization." For example if a company has 1,000,000 shares and is currently trading at $10 a share, their market capitalization is $10,000,000.[2]

In general, there are six types of capital, as shown in Figure 15.2.

15.3.1 How to Capitalize Your StartUp

The founders typically capitalize their "seed capital" activity by any of the methods shown here:

■ Sweat equity (bootleg)
■ SBIR (Small Business Innovation Research grants)
■ Friends, Family, Fools (the 3 Fs)

The Six Types of Capital

- Cash
- Equity
- Working capital (current assets – current liabilities)
- Capital assets (aka fixed assets)
- Invested capital (equity + long-term debt)
- All assets (aka everything you own)

Figure 15.2   Capitalization—The capitalization process from seed to IPO as a staged series of events.

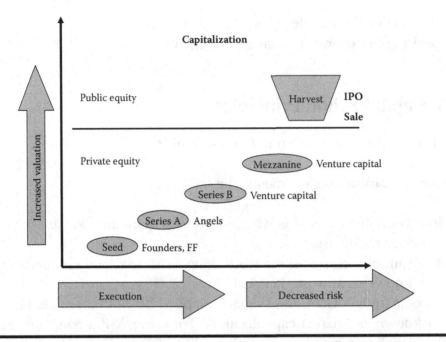

Figure 15.3　Capitalization at a glance—Capitalization "rounds" undertaken as the entrepreneur executes on the business plan.

- Banks
- Savings, second mortgage, etc.

Several additional capitalization rounds will be necessary. Figure 15.3 is a visual representation of the expected continuum, from seed to IPO.

Seed round. Earliest stage investment, usually undertaken by founders or Friends and Family (FF). Typically less than $50,000 in a medical device startup, or less than $1 million in a software company. Investment made to validating market size, developing a business model, and producing a proto-type product, thereby mitigating initial startup risks.

Series A round. The first substantial public financing event is usually undertaken by Angel investors. Typically, it is in the range of $250K to $500K for medical device companies and less than $5 million for a software com-pany. Money is used to build working prototype and set up a beta site for software companies.

Series B round. Often undertaken by venture capital, private equity, and hedge funds to scale up operations, build up teams, and start sales (software companies). Represents increased investor confidence in the executive team and technology following product-related milestones. There is increased investor acceptance and visibility.

Series C, D, E ... rounds. Additional capital needed for rapid sales growth (or start of animal trials for life science companies). The entire process can be viewed as shown in Figure 15.3.

15.4 Going Public

"Nobody goes there anymore; it's too crowded." —Yogi Berra

Going public can provide companies in the most competitive business environments with the capital required to exploit technologies and innovative products, fuel future growth, and provide funding for potential acquisitions.

Going public can be achieved in several ways: sale of common stock or preferred stock, public sale of debt, or a combination of these techniques. However, going public is a complex undertaking requiring the help of experts in the marketing of securities, legal, and accounting professions.

15.4.1 Advantages

There is great seductive allure to "taking your company public." Figure 15.4 summarizes some of the advantages of taking the company public, a process known as an Initial Public Offering (IPO).

Capitalization at a glance

B round	Staffing Infrastructure	$ 1–5 million
C round	Product development	$ 5–10 million
D round	Shipping Market share	$ 10–15 million

Figure 15.4   IPO advantages—The five major advantages to entrepreneurs by "taking the company public."

15.4.2 Disadvantages

There are several well-known counterbalancing pressures against going public. Among the major disadvantages are:

- High initial cost of going public
  - Underwriters' expenses
  - Road show expenses
  - Legal and accounting expenses

- High cost of compliance
  - Audit expenses
  - Burdensome regulations (Sarbanes-Oxley)
  - Periodic reporting requirements (10Qs, 10Ks).
  - Annual shareholder meetings

- Shareholder pressures
  - Increased quarterly earnings
  - Unrelenting pressure to increase quarterly share price
- Distant and unsympathetic shareholders

15.4.3 The Going Public Process

At first, management should seek the advice of attorneys who understand the complexities of securities law, independent public accountants experienced in public offerings, and investment bankers to advise on how to best position the company.

Established companies must present (1) audited financial statements for the past five years (or two years in the case of smaller public offerings), and (2) selected financial data going back five years. Auditing firms evaluate the company's internal report system and its accounting practices to ensure conformity with Securities and Exchange Commission (SEC) requirements.

The auditing firm should also determine whether the company's accounting system could produce the required financial statements for timely reporting once the company becomes publicly held to help underwriters, investors, or lenders gauge the company's current operations and its potential.

15.4.4 The Registration Process

The process of going public is governed by the SEC Act of 1933, which (1) requires that a registration statement be filed with the SEC *before* securities are offered for public sale, and (2) prohibits the sale of any securities until the registration is declared effective. It is virtually impossible for a company to sell its own securities directly to the public. Instead, this is handled by underwriting syndicates that maintain current investor lists and worldwide office networks. Table 15.1 summarizes the steps typically included in the registration process.

Table 15.1   Registration Process Steps

Underwriter Selection	Selection of lead underwriter that will form the syndicate that will buy the company securities and resell them to the public.
Company Evaluation	Underwriter evaluates the management, products, and markets to determine the marketability of the stock. Often, problems are encountered that must be corrected prior to the offering.
Letter of Intent	Preliminary and nonbinding agreement that estimates the size of the offering, the price range, as well as details of commissions and expense reimbursements.
"Quiet Period" (The period of time extended from the time a company files a registration statement with the SEC until SEC staff declares the registration statement effective.)	During the time the SEC "reviews" and "comments" on a potential registration, the SEC discourages initiation of publicity, issuance of forecasts related to revenues, income, or earnings per share, and the publication of opinions concerning valuations.
Drafting the Registration Statement	Prepare Form S-1 or Form SB-2 consisting of (a) the offering prospectus for eventual distribution to investors and (b) any supplemental information required by SEC.

(*Continued*)

Table 15.1 (*Continued*)

Registration Statement Filing	Official filing with SEC. At this point, the information contained becomes a matter of public record.
"Red Herring"	Before the Registration Statement becomes "effective," copies of the preliminary prospectus or "Red Herring" are distributed by the managing underwriter to the syndicate.
Due Diligence Meetings	The underwriters typically conduct an investigation to ensure that the registration statement does not contain misstatements or omissions of material facts using the Red Herring. At this point, company officials submit themselves to formal probing by underwriters, attorneys, and potential investors during Due Diligence meetings.
Compliance with **"Blue Sky"** Laws	Individual states may impose laws different from U.S. federal laws; these are called "Blue Sky" laws. The company must ensure that the registration complies with all applicable federal and state securities laws.
Deficiency Letter	After reviewing the registration statement, the SEC may issue a "Deficiency Letter" if issues need to be addressed or revised before the statement can become effective.
Offering Price, Size plus "Green Shoe"	Company and underwriters agree on the price and size of the offering based on preliminary interest by potential investors. The underwriters may also enter into an option known as a "Green Shoe" to acquire and resell an additional 15% of stock under a firm commitment to cover overallotments to customers.

Table 15.1 *(Continued)*

Comfort Letter	Auditors and attorneys issue a "Comfort Letter," a long and complicated document required by the underwriters. The document details the accountant's viewpoints and concerns on a variety of accounting issues and financial matters appearing in the prospectus.
Effective Registration Statement	When the SEC declares that the registration is "effective," the company and underwriters sign a formal and binding letter of agreement.
Publishing the Prospectus	The "Prospectus" is the part of the effective registration statement used as a selling document distributed by the underwriting syndicate for distribution to prospective investors. It is a lengthy document detailing information about the offering and the history of the company, including exhaustive financial information.
Closing Meeting	Final meeting to make the exchange of the securities for the proceeds of the offering. At this time, the underwriters purchase shares and begin reselling to the public.
Is There Life After an IPO?	For the company founders, a smaller slice of a rapidly expanding pizza is worth more than a larger slice of a smaller pizza. This is the entrepreneur's "pizza philosophy."

15.5 Strategic Alliances

"You have to break a few eggs to make omelets."

Strategic alliances represent a way for aggressive companies to pursue growth by broadening product lines, penetrating new markets, and stabilizing cyclical businesses despite limited resources.

A **strategic alliance** is an agreement between two or more parties to pursue a set of agreed upon objectives while remaining independent organizations. This form of cooperation lies between mergers/acquisitions and organic growth.[3] However, the exchange of managerial talent, resources, capabilities, and possibly an equity investment elevates the alliance beyond a mere contractual agreement. The three main characteristics of strategic alliances are summarized next:

■ Multifaceted, goal-oriented, long-term partnerships between two companies
■ Both risks and rewards are shared
■ Typically leads to long-term strategic relationship

15.6 The "Big Question"

Before embarking on any strategic alliance quest, management must answer the "Big Question":

■ Do we create organic sales growth vs. sales growth through acquisitions?

The firm can grow organically (by internal investment) or inorganically (by strategic alliances, i.e., cooperative ventures, joint ventures, joint ownership, or mergers and acquisitions). One theoretical way of approaching the "Big Question" is to look at the continuum of strategic partnership interdependence as shown in Figure 15.5.

When two or more companies combined participate in a project, it is a **cooperative venture**. This participation can be in the form of sharing financial or technical resources for mutual benefit.

A **joint venture** (JV) creates a separate entity in which both firms invest. The JV agreement specifies investment rights, operational responsibilities, voting control, exit alternatives, and the allocation of risks and rewards. The entity could be a division or an entirely new business established for the venture.

In a **joint ownership** alliance, the parties agree to long-term licensing agreement, co-marketing agreements, co-development agreements, joint

---

IPO advantages
■ Improved financial condition
■ Using stock for acquisitions
■ Using stock as employee incentive
■ Investor liquidity (exit)
■ Enhanced company visibility

Figure 15.5 Strategic alliance option—Strategic alliances may be a profitable way of growing the company rapidly.

purchasing agreements, or long-term supply or toll agreements, with each party owning 50% of the intellectual property plus other non-tangible assets.

A **merger & acquisition** (M&A) is a general term used to refer to the consolidation of two independent companies to form an entirely new company, while an acquisition is the purchase of one company by another in which no new company is formed.

15.6.1 Drivers of M&A Activities

"You name the price. I'll name the terms."

Historically, M&A presents management with the following hard-to-resist list of opportunities:

Geographic expansion
Market leadership position
Broadened intellectual property portfolio
Becoming a larger company
Manufacturing and distribution synergies
Immediate increase in infrastructure
Broader product offering

Every M&A transaction has its own set of unique reasons for combining the two companies. From management's standpoint, the underlying principle behind an M&A transaction is deceptively simple: $2 + 2 = 5$. The value of the acquirer is $2 billion and the value of the acquired is $2 billion, but when we merge the two companies, the market values the new entity as $5 billion. This is classical "synergy."

In addition to perceived synergies, there are real strategic drivers to M&A activities. Figure 15.6 summarizes the strategic picture.

15.6.2 Advantages of Strategic Alliances

Traditionally, inorganic growth can be achieved by the judicious use of strategic alliances. **Strategic alliances** are multifaceted, goal-oriented, long-term partnerships between two companies. In a strategic alliance, both risks and rewards are shared and typically lead to long-term strategic benefits for both partners, as summarized here:

■ Adding value to products
■ Improving market access

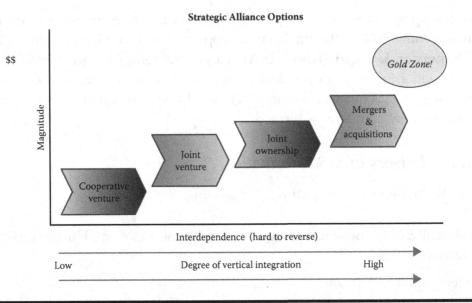

Figure 15.6   M&A: strategic drivers—Three compelling reasons for undertaking strategic alliances.

- Strengthening operations
- Adding technological strength
- Enhancing strategic growth
- Enhancing organizational skills
- Building financial strength

In addition, strategic alliances display other advantages, such as those seen in Figure 15.7.

Figure 15.7   Benefits of strategic alliances—Summary of benefits of strategic alliances.

Table 15.2 Mergers and Acquisitions by the Numbers [4]

Short-term stock price	Seller's price increases, whereas buyer's stock price decreases
Premium over pre-merger price = 38%	Announcement price reaction: Target = +16% Acquirer = −1%
Long-term performance	Over 80% lost market value in the first two years post-transaction

15.6.3 Pitfalls of Strategic Alliances

Buying all or part of a business is one of the most complex strategic moves a company can make. Despite the number and size of headline-making deals in the popular press, research indicates that the success rate of strategic alliances is very low. Short-term stock price results can be summarized as shown in Table 15.2 (see also Figure 15.8).

15.6.4 Best Practices

There are two schools of thought regarding mergers, acquisitions, and strategic alliances. Practitioners of the first school argue that transactions are accomplished by the hubris of the two respective executives, with details

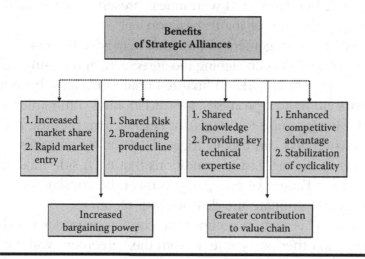

Figure 15.8   Pitfalls of strategic alliances—For every advantage, there is a corresponding disadvantage of strategic alliances.

negotiated later by the respective company specialists. The second school of thought recognizes the intricacies of the transaction and develops a systematic methodology, thus increasing the likelihood of long-term success.

The establishment of a methodology for analyzing a potential M&A is the difference between an amateur and an experienced buyer. Table 15.3 presents a summary of best practices to be followed (see also Figure 15.9).

At closing, a definitive agreement is reached. In contrast to the Letter of Intent, which is non-binding, the final agreement is definite; that is, it describes all the necessary details relevant to consummating the deal, and is a legally binding contract, subject to preconditions, such as shareholder approval.

The definitive agreement is a risk management device focused only on the completion of the transaction, and contains a number of elements in common, including these:

1. Parties to the deal. Specifies players and their roles.
2. Recitals. Specifies what the parties wish to accomplish, and is easily identified by clauses that begin with "Whereas."
3. Definition of terms. Mutually agreed understanding of the terminology contained in the agreement.
4. Description of transaction. Purchase or sale of assets or equity, or merger. Describes exactly what is to be exchanged, by whom, and when.
5. Representations and warranties. Enumerates mechanisms by which the two sides disclose information about each other. A **representation** is a statement of fact; a **warranty** is a commitment that a fact is or will be true. Together, "reps and warranties" present a snapshot of the target and buyer at the time of the transaction.
6. Covenants. The management of risks that may arise because of the behavior of the parties between signing the agreement and closing the transaction. (Closing is not considered finalized until funds have been transferred.) **Covenants** are mutual promises, forward-looking commitments. They are affirmative (we promise to do this), or negative (we promise not to do that). Breach of covenants can usually trigger litigation for damages.
7. Conditions to closing. List of conditions that each side must fulfill in order to close. Failure of one party to meet the conditions permits the other party to terminate the deal without recourse.
8. Termination. This section outlines the conditions under which one party will allow the other party to exit from the agreement without penalty.
9. Indemnifications. Damage payments in the event of losses discovered after closing has occurred, or even breach of provisions in the agreement.

Table 15.3 Best Practices for M&A Negotiations

Common Mistakes	Common Solutions
Unrealistic timetables and expectations	Select an acquisitions team experienced in cost/benefit time allocation
Inexperienced M&A negotiation team	Establish an acquisitions team with defined responsibilities and authority
Lack of structured transaction process	Clearly define team member roles Be prepared to "walk away" from a bad deal
Disproportionate time spent on minor issues	Focus on outcomes, not activities Set "drop dead" dates
Incomplete or irrelevant information Inadequate or nonexistent due diligence	Obtain corroborating data Engage industry experts
Inadequate sensitivity analysis	Perform financial and commercial sensitivity analyses Set minimums
Overlooking integration issues	Thoroughly assess the impact of "culture clashes" Can value be created?
Poor negotiating techniques Naïve and inexperienced negotiators	Pre-plan negotiation strategies, tactics, and strategic objectives Decide on deal breakers ahead of time
No meeting transcriptions or minutes Communications failures	Include a "secretary" as note taker Debrief other party on issues discussed and agreements reached
Long time to reach a Term Sheet agreement. Deals have a life. A lengthy negotiation for a Term Sheet is an early warning sign of impending impasse.	This is the most significant document in the early stages. It should list price, form of payment, deal structure and management issues. The Letter of Intent will follow the overall principles contained in the Term Sheet.
Over-reliance on a Letter of Intent by seller. It is merely "an agreement to agree."	Understanding that a Letter of Intent is a non-binding agreement; while it is crucially important, an agreement to agree is not an agreement

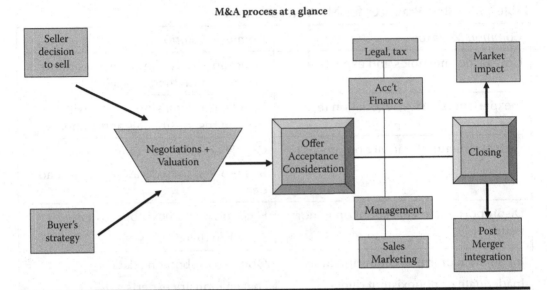

Figure 15.9 M&A process at a glance—Visual representation of the M&A process from start of negotiations to post-closing activities.

15.7 Mergers and Acquisitions

Mergers and acquisitions (abbreviated **M&A**) is an aspect of corporate strategy, corporate finance and management dealing with the buying, selling, dividing, and combining of different companies and similar entities that can help an enterprise grow rapidly in its sector or location of origin, or a new field or new location, without creating a subsidiary, other child entity, or using a joint venture.[5]

15.7.1 Mergers

Theoretically, a **merger** happens when two firms agree to go forward as a single new company rather than remain separately owned and operated. This kind of action is more precisely referred to as a "merger of equals." In practice, however, actual mergers of equals do not happen very often. Usually, one company will buy another and, as part of the deal's terms, simply allow the acquired firm to proclaim that the action is a merger of equals, even if it is technically an acquisition.[6]

15.7.2 Acquisitions

Practically speaking, an **acquisition** is the process through which one company completely takes over the controlling interest of another company. Such controlling interest may be 100%, or nearly 100%, of the assets or ownership equity of the acquired entity. An "acquisition" usually refers to a purchase of a smaller firm by a larger one (Figure 15.10).

There are two sets of stockholders affected by a merger or acquisition: those of the firm being acquired and those of the firm doing the acquiring.

15.7.3 The Challenge of M&A

There are many examples of successful M&A (and many more failures). The most spectacular M&A failure of all times was the acquisition of Time Warner by AOL. In January 2000, America Online merged with Time Warner in a deal valued at a stunning $350 billion. It was then, and is now, the largest merger (and eventual de-merger) in American business history.

Due to the larger market capitalization of AOL, they would own 55% of the new company while Time Warner shareholders owned only 45%. So in actual practice AOL had acquired Time Warner, even though AOL had far less assets and revenues.[7]

What was the perception of the average stockholder when AOL offered such a large premium over market price to purchase Time Warner? The average AOL stockholder wondered whether the acquisition was being made in their interest or in the interest of management. Some important lessons have

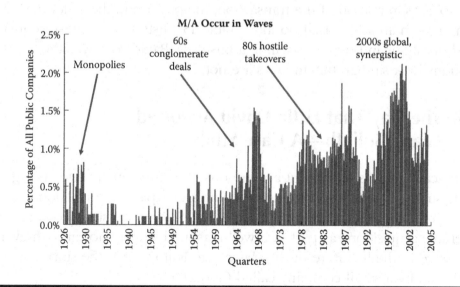

Figure 15.10   M/A occur in waves—Due to economic/regulatory changes, M/A activity occurs in waves.

been learned from the AOL/Time Warner failure. The following five criteria have been offered as success principles for M&A/strategic alliances:[8]

- Compatible strategy and culture
- No overpayment
- Comparable contribution
- Compatible strengths
- No conflict of interest

So, what is an M&A success and what is an M&A failure? Is it a matter of luck, perception, or timing? The author recommends a quantitative measure, based on Return on Invested Capital (ROIC) and Weighted Average Cost of Capital (WACC) as shown next:

$$\text{Failure} \quad \text{ROIC} < \text{WACC}$$

$$\text{Success} \quad \text{ROIC} > \text{WACC}$$

where

ROIC = earnings before interest and taxes divided by (invested capital equity + long-term liabilities)

WACC = Sum of after-tax of debt and equity, multiplied by their respective percentage of capital structure

15.7.4 The Enduring Questions

Prior to consummation of the transaction, the buyer and the seller find themselves in an adversarial position. Table 15.4 lists the ten most frequently asked questions in buying or selling a business. Paradoxically, although the questions look similar, the answers are not.

15.8 The Day That Little David Acquired Giant Goliath—A Case Study

When academicians talk about acquisitions, they are usually referencing events such as: "Johnson & Johnson acquires Company X for billions," or "Apple Inc. acquires Company Y for billions." Seldom do small companies undertake acquisitions as part of a well-reasoned strategy for growth even mentioned in the literature or the classroom. Following is the story of an acquisition by a small company called CardioTech.

By 2002, CardioTech (a company founded by Dr. Michael Szycher and publically traded in the AMEX), had decided to grow inorganically by

Table 15.4 The Most Frequently Asked Questions by Sellers and Buyers

Seller	Buyer
Why is their offering price so low?	Why is their asking price so high?
Are they bottom fishing?	Are they serious about selling?
Is it okay for us to shop around?	Should this be a no-shop negotiations?
Why are they interested in us?	What do we really know about them?
Do we really need money now? Is this transaction for cash or shares?	What will they do with our money? What are their future financial demands?
Who should be part of our negotiation team?	Are we really negotiating with the decision makers?
How much independence do we retain post-transaction?	How do we integrate them into our winning culture?
Could we do better on our own?	Do we really need them?
What happens if the deal collapses? How vulnerable do we remain?	What if we find deal-killers during due diligence?
What happens if our CEO is run over by the mythical train during negotiations?	How indispensable is their CEO? Do they have a succession strategy?

opportunistically pursuing an accretive acquisition strategy. To reduce any managerial impulsive and emotional decisions, CardioTech pre-established a systematic, quantitative acquisition criteria matrix, as shown in the following:

- Balance Sheet Criteria
  - Debt-to-equity ratio
  - Working capital
  - Liquidity requirements
- Performance Criteria
  - Return on equity
  - Return on invested capital
  - Return on assets
  - Earnings growth rates
- Business Characteristics Criteria
  - Relative market position
  - Market share
  - Competitive strengths
  - Product line

- Cash generation
- Price and cost structure
- Geographical reach
■ Market/Industry Characteristics Criteria
- Barriers to entry
- Governmental regulations
- Industry growth
- Company growth cycle
■ Investment Criteria
- Quality of earnings
- History of labor relations
- Labor intensive or technology intensive
■ Intellectual property portfolio

This case study exemplifies the unusual occurrence when CardioTech (little David with $3 million in sales) acquired Gish Biomedical (giant Goliath with $17 million in sales). Please note this was not a merger; it was an acquisition. At the end of the transaction, CardioTech owned 100% of Gish. That is, Gish became an operating, fully owned subsidiary of CardioTech.

After establishing the above criteria for the acquisition, CardioTech started the search for an appropriate target. The following paragraphs enumerate the history behind the acquisition.

15.8.1 The Acquisition

On April 7, 2003, CardioTech completed the acquisition of Gish Biomedical. Gish shareholders were issued 1.3422 shares of CardioTech stock for each of their shares of Gish common stock. Gish Biomedical, Inc., a California corporation, was founded in 1976 to design, produce, and market innovative specialty surgical devices. Gish developed and marketed its innovative and unique devices for various applications within the medical community. Gish operated in the medical devices segment—the manufacture of medical devices—which were marketed through direct sales representatives and distributors domestically and through international distributors. All of Gish's products were single-use disposable products or had a disposable component. Gish's primary markets included products for use in cardiopulmonary bypass surgery, myocardial management, infusion therapy, and postoperative blood salvage.

Prior to the transaction, Gish had been experiencing a steady 10% decline in annual sales for the previous 5 years. This followed the overall downward trend in the number of coronary artery bypass (CABG) procedures, which started in 1996. In 1996, a revolutionary and disruptive technology made its debut: coronary artery stents, implanted via a minimally invasive percutaneous procedure. There had been a dramatic drop in the rate of heart bypass procedures performed in the U.S. over the last decade, even though more hospitals were offering open-heart surgery.

15.8.2 Open Heart Procedures

Since its introduction in the mid-1960s, CABG had become the most frequently performed surgical heart procedure. CABG performed miracles for the millions of patients suffering from atherosclerotic coronary artery disease. However, CABG required opening the patient's chest in a 4- to 6-hour procedure, kept patients in the hospital for nearly a week, and had several troubling side effects.

In contrast, coronary stents were implanted by accessing an artery in the groin, most procedures required between 45 and 60 minutes, and the patients were discharged on either the same day or the day subsequent to the procedure. Stents revolutionized coronary artery revascularization, displacing cardiopulmonary CABG procedures nearly overnight. This can be seen in Figure 15.11, which shows the dramatic decline of CABG procedures starting in 1996, the year stents were introduced into the interventional cardiology marketplace.

Gish's sales declined in tandem with the rest of the decline in cardiopulmonary bypass procedures. The entire industry was in deep decline, but Gish suffered the most by being the smallest participant. Other industry players, such as Medtronic, Terumo, and Sorin were much larger companies, and were better able to withstand steep sales declines in one of their divisions. However, Gish was heavily dependent on CABG procedures, and thus proportionately most susceptible to market declines. By 2002, Gish was nearly bankrupt on yearly sales of approximately $17 million per year. Moreover, the future looked very bleak as seen in Figure 15.12.

15.8.3 CardioTech as the White Knight

With bad news like this, why would CardioTech want to acquire Gish? For several reasons:

- Gish management payroll was heavily bloated with highly paid executives who contributed very little.
- Gish required leadership, not management skills.

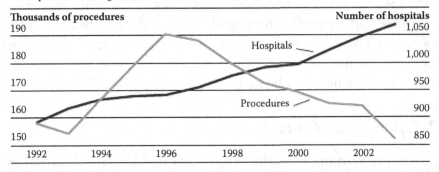

Annual Number of coronary Artery Bypass Graft (CABG) Procedures and Number of Hospitals Performing CABG in The United States, 1992–2003

Figure 15.11 Decline in CABG procedures—Due to the introduction of minimally invasive stent procedures in 1996, the number of CBG procedures suffered a huge decline.

- At $17 million per year of sales, it *should* be operationally profitable.
- Gish needed innovation. This took the form of a novel antithrombotic coating, as required by industry.
- Gish needed to expand into the Asian market.
- Gish agreed to be acquired with CardioTech stock.

If CardioTech was able to consummate this acquisition and make the necessary management and technical changes (remember culture?), then CardioTech would emerge with $21 million in sales and become

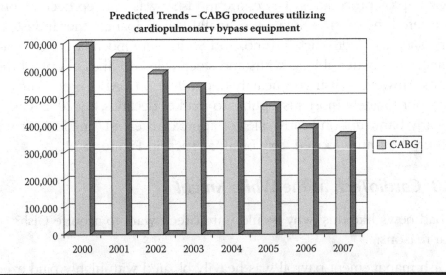

Figure 15.12 Predicted trends—With the decline in CABG surgeries, the sales of cardiopulmonary bypass equipment declined correspondingly.

operationally profitable. The technical term for a financial transaction of this type is "accretive." **Accretive** is an acquisition that increases a publicly traded company's earnings per share. An accretive acquisition occurs when the price-earnings ratio of the acquiring firm is greater than that of the target firm. This means that the target firm's earnings are likely strong; this is often seen as a good investment. An accretive acquisition usually results in a higher share price for the combined company.[9]

CardioTech's founder and CEO, Dr. Michael Szycher, was an old hand at accretive acquisitions, having been "trained" at Thermo Electron in many accretive acquisitions. Together with his CFO David Volpe, and a terrific law firm, Ellenoff Grossman & Schole LLP of New York, they set out "to make it happen." In the meantime, the joke among Wall Street bankers was that Dr. Szycher "never saw an acquisition he did not like."

15.8.4 Risk Factors SWOT Analysis of the Gish Acquisition

The CardioTech acquisition team developed a process for potential acquisition and SWOT analysis, shown in Figures 15.13 and 15.14.

15.8.4.1 Strengths

■ Gish enjoyed a well-known brand name in the industry, and was known for high-quality products.

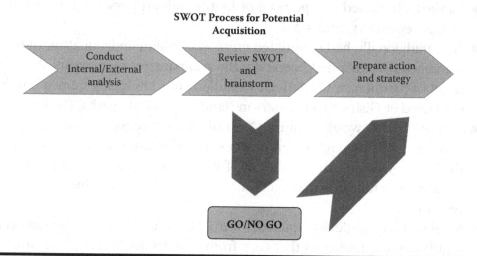

Figure 15.13 SWOT process—The go/no go decision tree for any potential acquisition.

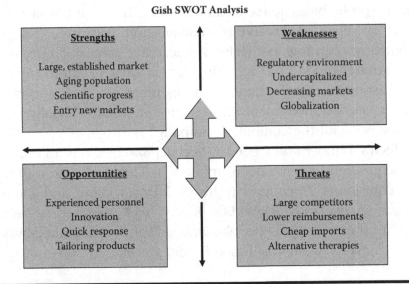

Gish SWOT Analysis

Strengths	Weaknesses
Large, established market	Regulatory environment
Aging population	Undercapitalized
Scientific progress	Decreasing markets
Entry new markets	Globalization
Opportunities	**Threats**
Experienced personnel	Large competitors
Innovation	Lower reimbursements
Quick response	Cheap imports
Tailoring products	Alternative therapies

Figure 15.14   Gish SWOT analysis—Strategic analysis for the potential acquisition of Gish, Inc.

■ International net revenues accounted for only 19% of Gish's total net sales in fiscal 2002 and 2001. CardioTech planned a major push to increase international sales revenues to 25% of net sales within 2 years of acquisition, using native distributors.

■ CardioTech gained a huge FDA-approved clean room facility, and a large warehouse plus a distribution facility in western U.S.

■ CardioTech gained a large pool of highly trained personnel in manufacturing, regulatory, and sales.

■ Gish traditionally had committed employees. Many individuals in the manufacturing area had been with the company for many years. As a result of the merger, CardioTech would acquire 116 highly trained employees based at Gish's headquarters in Rancho Santa Margarita, California.

■ Future growth would require CardioTech to successfully hire, train, motivate, and manage its employees. In addition, CardioTech's continued growth and the evolution of its business plan would require significant additional management and technical and administrative resources.

■ Sales of medical devices outside of the U.S. are subject to international regulatory requirements that vary from country to country. The time required to obtain approval for sales internationally is much shorter than that required for FDA clearance or approval, and the requirements may differ. Gish entered into distribution agreements for the foreign distribution of its products. These agreements are an encouraging sign of possible new sales.

15.8.4.2 Weaknesses

- Limited financial resources. Several large companies offered devices that directly competed with devices manufactured by Gish, including Jostra-Bentley, COBE Cardiovascular, a division of Sorin Biomedica, Terumo, Medtronic, Inc., and Stryker Surgical. Most of Gish's competitors have longer operating histories and significantly greater financial, technical, research, marketing, sales, distribution, and other resources than Gish has. In addition, Gish's competitors had greater name recognition than Gish and frequently offered discounts as a competitive tactic.
- Liabilities. Gish's acquisition exposed CardioTech to potential product liability and other liability risks inherent in the development, testing, manufacturing, and direct marketing of medical products.
- Limited international sales. International net revenues accounted for approximately 19% of Gish's total net sales in fiscal 2002 and 2001. International sales are subject to a number of inherent risks, including the impact of possible recessionary environments in economies outside the U.S., unexpected changes in regulatory requirements, and fluctuations in exchange rates of local currencies in markets where Gish sells its products. While Gish denominated all of its international sales in U.S. dollars, a relative strengthening in the U.S. dollar would increase the effective cost of the company's products to international customers.
- Organizational disruption. The potential disruption of the combined organizations' ongoing business and distraction of its management from the day-to-day operations of the combined organization.
- Integration of cultures. The combined organization may not succeed in addressing these risks or any other problems encountered in connection with the merger. The inability to integrate the operations, technology, and personnel of CardioTech and Gish successfully, or any significant delay in achieving integration, would have an adverse effect on the combined organization after the merger and, as a result, on the corporate profitability and market price of CardioTech common stock.
- Impact on Gish employees. Because of Gish's change in ownership, current and prospective Gish employees would experience uncertainty about their future roles within the combined organization. This

uncertainty would adversely affect the ability of Gish to attract and retain key management, sales, marketing, and technical personnel. In addition, in connection with the merger, certain employees are entitled to acceleration of vesting of stock options, which would adversely affect the ability of CardioTech to retain such employees following the acquisition. Any failure to attract and retain key personnel would have a material adverse effect on the combined businesses of Gish and CardioTech.

15.8.4.3 Opportunities

■ The manufacture and sale of medical devices, including products currently sold by Gish and Gish's other potential products, are subject to extensive regulation by numerous governmental authorities in the U.S., principally the FDA, and corresponding state agencies, such as the California Department of Health Services (CDHS). In order for Gish to market its products for clinical use in the U.S., Gish must obtain clearance from the FDA of a 510(k) pre-market notification or approval of a more extensive submission known as a pre-market approval (PMA) application. CardioTech had experience in obtaining 510(k) approvals from FDA.

■ Gish did not have heparin coatings for its product line. By 2003, the industry required all blood-contacting surfaces to be coated with an antithrombotic surface. CardioTech had expertise in this area, and could be of great assistance in developing and introducing coated products.

■ CardioTech planned a subsequent acquisition of an established and highly profitable medical device company in China, thus opening the Chinese market to Gish's products.

15.8.4.4 Threats

■ Introduction of substitute products. The entry of minimally invasive stents as a disruptive new procedure resulted in cataclysmic reductions in coronary-bypass open heart surgeries. The industry was suddenly faced with a supply overcapacity that became more produced year after year.

■ Downward pricing pressures. Gish was facing increasing downward pricing pressures from domestic and international competitors, especially from competitors specializing in the cardiovascular surgery products market. Governmental reimbursements were being steadily eroded, as cardiopulmonary products were becoming more "commoditized."

- Political, economic, and regulatory influences were subjecting the healthcare industry in the U.S. to fundamental change. Potential reforms proposed have included mandated basic healthcare benefits, controls on healthcare spending through limitations on the growth of private health insurance premiums and Medicare and Medicaid spending, the creation of large insurance purchasing groups, and fundamental changes in the healthcare delivery system. The CardioTech/Gish combined company must be prepared to address healthcare issues in a systematic and orderly fashion.

15.8.4.5 The Stock-for-Stock Transaction

Gish Selling Security Holder	Number of Shares of CardioTech Common Stock Expected to be Beneficially Owned Prior to the Offering	Number of Shares of CardioTech Common Stock Offered
Asset Value Fund Limited Partnership (1)	792,435	792,435
Craig Corporation (through its controlling interest in Citadel Holding Corporation)	783,711	783,711
Gish CEO	527,887	138,649
Investor 1	186,821	180,110
Investor 2	12,319	5,608
Total	2,303,173	1,900,513

15.8.5 An Accretive Acquisition

An accretive acquisition occurs when (1) the acquiring company's earnings per share increase, thus creating shareholder value, or (2) when the price-earnings ratio of the acquirer is greater than the target.

By either measure, CardioTech's acquisition of Gish was accretive, as shown below:

	2001	2002	2003	2004	2005	2006
CardioTech	1543	3220	3394	21,799	21,841	22,381
Gish	18,017	16,410	17,584			
	19,560	19,630	20,978	21,799	21,841	22,381
Shares Outstanding @ 3/31			9,126,889	17,654,854	19,258,983	19,796,833
Stock Price @ 3/31			$1.03	$5.15	$1.90	$2.78
Market Capitalization @ 3/31			$9,400,696	$90,922,498	$36,592,068	$55,035,196

15.9 Sample Prospectus

"A Prospectus is Latin for close your eyes and open your wallet."

As filed with the Securities and Exchange Commission on November 26, 2003

SECURITIES AND EXCHANGE COMMISSION
Washington, D.C. 20549

FORM S-3

REGISTRATION STATEMENT
UNDER THE SECURITIES ACT OF 1933

CARDIOTECH INTERNATIONAL, INC.

(Exact name of registrant as specified in its charter)

MASSACHUSETTS	2834	04-3186647
(State or other jurisdiction of incorporation or organization)	(Primary Standard Industrial Classification Code Number)	(I.R.S. Employer Identification No.)

78-E Olympia Ave.
WOBURN, MASSACHUSETTS 01801
(781) 933-4772

(Address, including zip code, and telephone number, including area code, of registrant's principal executive offices)

DR. MICHAEL SZYCHER
CHAIRMAN AND CHIEF EXECUTIVE OFFICER
CARDIOTECH INTERNATIONAL, INC.
78-E Olympia Ave.
WOBURN, MASSACHUSETTS 01801
(781) 933-4772

(Name, address, including zip code, and telephone number, including area code, of agent for service)

Copies to:

DAVID SELENGUT, ESQ.
ELLENOFF GROSSMAN & SCHOLE LLP
370 Lexington Avenue
NEW YORK, NY 10017
(212) 370-1300

Approximate date of commencement of proposed sale to the public: As soon as practicable from time to time after the Registration Statement becomes effective.

If any of the securities being registered on this Form are to be offered on a delayed or continuous basis pursuant to Rule 415 under the Securities Act of 1933, check the following box. ☐

If this Form is filed to register additional securities for an offering pursuant to Rule 462(b) under the Securities Act, please check the following box and list the Securities Act registration statement number of the earlier effective registration statement for the same offering. ☐

If this Form is a post-effective amendment filed pursuant to Rule 462(c) under the Securities Act, check the following box and list the Securities Act registration statement number of the earlier effective registration statement for the same offering. ☐

If this Form is a post-effective amendment filed pursuant to Rule 462(d) under the Securities Act, check the following box and list the Securities Act registration statement number of the earlier effective registration statement for the same offering. □

If delivery of the prospectus is expected to be made pursuant to Rule 434, please check the following box. □

CALCULATION OF REGISTRATION FEE

Title of each class of securities to be registered	Amount to be registered(1)	Proposed maximum offering price per share(2)	Proposed maximum aggregate offering price(2)	Amount of registration fee
Common stock, $.01 par value	410,000	$5.70	$2,337,000	$189.06

(1) We are registering a total of 410,000 shares of our common stock issuable upon exercise of warrants issued to certain stockholders at various exercise prices.

(2) Estimated solely for the purposes of determining the registration fee. In accordance with Rule 457(c) under the Securities Act of 1933, as amended, the above calculation is based on the approximate average of the high and low prices reported on the American Stock Exchange on November 20, 2003.

The registrant hereby amends this Registration Statement on such date or dates as may be necessary to delay its effective date until the registrant shall file a further amendment which specifically states that this Registration Statement shall thereafter become effective in accordance with Section 8(a) of the Securities Act of 1933 or until the Registration Statement shall become effective on such date as the Commission, acting pursuant to said Section 8(a), may determine.

PROSPECTUS

Registration No. [ ]

410,000 SHARES

of

CARDIOTECH INTERNATIONAL, INC.

COMMON STOCK

(Par Value $.01 per share)

This prospectus relates to the offering for resale of CardioTech International, Inc. common stock. In this prospectus, the terms "CardioTech," "we," or "us" will each refer to CardioTech International, Inc. The shares of common stock being offered were issued to the selling security holders upon exercise of certain warrants, which were issued in connection with a private placement of units, each unit consisting of one share of common stock and one warrant to purchase one share of common stock. Sales of the units were made pursuant to Unit Purchase Agreements between CardioTech and each of the unit purchasers, dated as of November 9, 1998 and December 15, 1998. The private placement terminated on December 15, 1998. Fechtor, Detwiler & Co., Inc. acted as placement agent for the offering. This prospectus will be used by selling security holders to resell their shares of common stock. We will not receive any proceeds from sales by the selling security holders.

Our common stock is traded on the American Stock Exchange under the symbol "CTE." On November 20, 2003, the closing price of our common stock on the American Stock Exchange was $5.70 per share.

We are a Massachusetts corporation. Our principal offices are located at 78-E Olympia Ave., Woburn, Massachusetts 01801 and our telephone number is (781) 933-4772.

Investing in our common stock involves risks. Please read carefully the section entitled "risk factors" beginning on page 4.

Neither the Securities and Exchange Commission nor any state securities commission has approved or disapproved of these securities or determined if this prospectus is truthful or complete. Any representation to the contrary is a criminal offense.

TABLE OF CONTENTS

About This Prospectus

The Company

Risk Factors

Use of Proceeds

Selling Security Holders

Plan of Distribution

Legal Matters

Experts

Where You Can Find More Information

Special Note Regarding Forward-Looking Information

You should rely only on the information contained or incorporated by reference in this prospectus. We have not authorized anyone to provide you with different information. You should not assume that the information contained in this prospectus is accurate as of any date other than the date on the front of this prospectus.

Information on any CardioTech internet web site or the web site of any subsidiary of CardioTech is not part of this document and you should not rely on that information in deciding whether to purchase shares of CardioTech common stock, unless that information is also in this document or in a document that is incorporated by reference in this document.

ABOUT THIS PROSPECTUS

This prospectus is part of a registration statement that we filed with the Securities and Exchange Commission (the "SEC") on Form S-3. Under the S-3, the selling security holders may, from time to time, sell the securities described in this prospectus. Each time a selling security holder sells securities, the selling security holder is required to provide you with this prospectus and, in certain cases, a prospectus supplement containing specific information about the selling security holder and the terms of the securities being offered. Any prospectus supplement may add, update, or change information in this prospectus. If there is any inconsistency between the information in this prospectus and any prospectus supplement, you should rely on the information in that prospectus supplement. Before deciding to purchase the securities being offered by the selling security holders, you should read both this prospectus and any prospectus supplement together with all additional information described under "Where You Can Find More Information."

THE COMPANY

CardioTech International, Inc., a Massachusetts corporation, operating since 1993, specializes in developing and manufacturing small-bore vascular grafts, or synthetic blood vessels, made of ChronoFlex, a family of polyurethanes that has been demonstrated to be biocompatible and non-toxic. Vascular grafts are used to replace, bypass, or provide a new lining or arterial wall for occluded, damaged, dilated or severely diseased arteries. CardioTech is developing a layered microporous coronary artery bypass graft trade named CardioPass. Through Gish Biomedical, Inc., its wholly-owned, subsidiary ("GISH"), CardioTech designs, produces and markets innovative specialty surgical devices. Gish's products are single-use disposable products or have a disposable component. Gish's products include devices for use in cardiac surgery, myocardial management, infusion therapy, and post-operative blood salvage.

RISK FACTORS

In addition to the other information contained in, or incorporated by reference into, this prospectus, potential investors should consider the following risk factors in evaluating whether to invest in our common stock. Additional risks and uncertainties not currently known to us, or that are not currently believed to be important to potential investors, if they materialize, may also adversely affect an investment in our common stock.

RISKS RELATED TO OUR OPERATING RESULTS

We Have Incurred Substantial Operating Losses and Risk Never Generating Any Meaningful Revenue or Earning Any Profits

We had sales of $3,394,000 for the year ended March 31, 2003 and we incurred losses aggregating $963,000 in the same period. An investor in our shares must assume the risk that we will never be profitable. None of our coronary artery graft products and technologies has ever been utilized on a commercial basis. Our ability to generate enough revenues to achieve profits will depend on a variety of factors, many of which are outside our control, including:

- the size of the market for our potential products,
- competition and other solutions,
- the extent of patent and intellectual property protection afforded to our products,
- the cost and availability of raw material and intermediate component supplies,
- changes in governmental (including foreign governmental) initiatives and requirements,
- changes in domestic and foreign regulatory requirements,
- the costs associated with equipment development, repair and maintenance, and
- the ability to manufacture and deliver products at prices that exceed our costs.

The Medical Device Industry Is Cyclical, and an Industry Downturn Could Adversely Affect Our Operating Results

Business conditions in the medical device industry have rapidly changed between periods of strong and weak demand. The industry is characterized by:

- periods of overcapacity and production shortages;
- cyclical demand for products;
- changes in product mix in response to changes in demand;
- variations in manufacturing costs and yields;
- rapid technological change and the introduction of new products by customers;
- price erosion; and
- expenditures for product development.

These factors could harm CardioTech's business and cause its operating results to suffer.

Our Gish Subsidiary Is Dependent on a Few Critical Vendors

Our Gish subsidiary has various "sole source" vendors who supply key components for Gish's products. While Gish believes alternate supply sources could be developed, Gish could incur significant costs to obtain alternate components. The alternate components could also require regulatory approval, the denial or delay of which, could adversely affect Gish's ability to provide products to its customers.

Our Gish Subsidiary Faces a Risk of Declining Average Selling Prices for Its Products

Gish is currently facing and may continue to face increasing pricing pressures from its current and future competitors, especially from competitors in the cardiovascular surgery products market. As a result of such pressures, Gish may be forced to lower prices in order to maintain its market share. There can be no assurance that Gish will be able to maintain its market share in the cardiovascular surgery products market in the face of continuing pricing pressures. Any material reduction in the prices for Gish's products would negatively affect CardioTech's gross margin.

We Have Limited Manufacturing Experience in Arterial Grafts and Once Our Products Are Approved, We May Not Be Able to Manufacture Sufficient Quantities at an Acceptable Cost

We are still in the research and development phase of our arterial grafts. Accordingly, once our products are approved for commercial sale we will need to establish the capability to commercially manufacture our product(s) in accordance with FDA and other regulatory requirements. We have limited experience in establishing, supervising, and conducting commercial manufacturing. If we fail to adequately establish, supervise, and conduct all aspects of the manufacturing processes, we may not be able to commercialize our products. We do not presently own manufacturing facilities necessary to provide clinical or commercial quantities of our coronary artery bypass graft.

We presently rely on a third party contractor, LeMatre Vascular, Inc., to manufacture our coronary artery bypass grafts. This exposes us to the risk of not being able to directly oversee the production and quality of the manufacturing process. This contractor, may experience regulatory compliance difficulty, mechanical shut downs, employee strikes, or any other unforeseeable acts that may delay production.

RISKS RELATED TO THE MARKETS WE COMPETE IN

The Medical Device Industry Is Very Competitive

The medical device industry in general, and the market for products for use in cardiovascular surgery in particular, is intensely competitive. Clients have numerous choices in terms of firms that produce prototypes or test products for medical devices, and most of our competitors have longer operating histories and significantly greater financial, technical, research, marketing, sales, distribution and other resources than we do. There can be no assurance that our current competitors or potential future competitors will not succeed in developing or marketing technologies and products that are more effective or commercially attractive than those that have been and are being developed by us, or that such companies will not succeed in obtaining regulatory approval for, introducing or commercializing any such products prior to us. Any of the above competitive developments could have a material adverse effect on our business, financial condition and results of operations.

The Cardiothoracic Surgery Market Is Undergoing Constant Technological Change

The cardiothoracic surgery market for our products is characterized by:

- changing technologies;
- changing customer needs;
- frequent new product introductions and enhancements;
- increased integration with other functions; and
- product obsolescence.

To develop new products and designs for our cardiothoracic market, we must develop, gain access to and use leading technologies in a cost effective and timely manner and continue to expand our technical and design expertise. Failure to do so could cause us to lose our competitive position and seriously impact our future revenues.

New Product Development in the Medical Device Industry Is Both Costly and Labor Intensive With Very Low Success Rates for Successful Commercialization

Our success is dependent in part on the design and development of new products in the medical device industry. The product development process is time-consuming and costly, and there can be no assurance that product development will be successfully completed, that necessary regulatory clearances or approvals will be granted by the FDA on a timely basis, or at all, or that the potential products will achieve market acceptance. Failure to develop, obtain necessary regulatory clearances or approvals for, or successfully market potential new products could have a material adverse effect on our financial condition and results of operations.

RISKS RELATING TO OUR PRODUCTS AND TECHNOLOGY

There Is a Risk of a Market Withdrawal or Product Recall in the Medical Device Industry

Complex medical devices, such as many of our products, can experience performance problems in the field that require review and possible corrective action by the manufacturer. We periodically receive reports from users of our products relating to performance difficulties they have encountered. We expect that we will continue to receive customer reports regarding the performance and use of our products. Furthermore, there can be no assurance that component failures, manufacturing errors or design defects that could result in an unsafe condition or injury to the patient will not occur. If any such failures or defects were deemed serious, we could be required to withdraw or recall products, which could result in significant costs. There can be no assurance that market withdrawals or product recalls will not occur in the future. Any future product problems could result in market withdrawals or recalls of products, which could have a material adverse effect on our business, financial condition, and results of operations.

We Are Exposed to Product Liability, Clinical and Preclinical Liability Risks, Which Could Place a Substantial Financial Burden Upon Us, Should We Be Sued. We Maintain a Minimal Level of Product Liability Insurance Above and Beyond Our General Insurance Coverage.

The manufacture and sale of medical products entail significant risk of product liability claims. Such claims may be asserted against us. In addition, in our clinical trials, we may use medical products that our potential collaborators may develop, and the subsequent sale of these products by us or by our potential collaborators may cause us to bear a portion of or all product liability risks associated with those products. A successful liability claim or series of claims brought against us could have a material adverse effect on our business, financial condition and results of operations.

We do not currently have any product liability insurance relating to clinical trials. We may not be able to obtain or maintain adequate product liability insurance on acceptable terms, if at all. Furthermore, our current and potential future may not be willing to indemnify us against these types of liabilities and may not themselves be sufficiently insured or have a net worth sufficient to satisfy any product liability claims.

We currently maintain product liability insurance for our Gish subsidiary, but there can be no assurance that Gish's existing annual insurance coverage limits of $3 million per occurrence and $3 million in the aggregate will be adequate to protect us from any liabilities that we might incur in connection with the clinical trials or sales of Gish's products. In addition, we may require increased product liability coverage if and when our products under development are successfully commercialized. Such insurance is expensive and in the future may not be available to us on acceptable terms, or at all. A successful product liability claim or series of claims brought against us or one of our subsidiaries in excess of its insurance coverage could have a material adverse effect on our business, financial condition and results of operations.

We May Not Be Able to Protect Our Intellectual Property Rights Adequately

Our ability to compete is dependent upon our ability to protect our intellectual property rights. We rely on a combination of patents, trademarks, copyrights, trade secrets, confidentiality procedures and non-disclosure and licensing arrangements to protect our intellectual property rights. Despite these efforts, we cannot be certain that the steps we take to protect our proprietary information will be adequate to prevent misappropriation of our technology, or that our competitors will not independently develop technology that is substantially similar or superior to our technology. More specifically, we cannot assure you that any future applications will be approved, or that any issued patents will provide us with any competitive advantages or will not be challenged by third parties. We cannot assure you that, if challenged, our patents will be found to be valid or enforceable, or that the patents of others will not have an adverse effect on our ability to do business. Furthermore, others may independently develop similar products or processes, duplicate our products or processes, or design their own products around any patents that may be issued to us.

We Could Be Harmed by Litigation Involving Patents and Other Intellectual Property Rights

None of our patents or other intellectual property rights has been successfully challenged to date. However, in the future, we could be accused of infringing the intellectual property rights of other third parties. We also have certain indemnification obligations to certain customers with respect to the infringement of third party intellectual property rights by our products. No assurance can be provided that any future infringement claims by third parties or claims for indemnification by our customers or end users of our products resulting from infringement claims will not be asserted or that assertions of infringement, if proven to be true, will not harm our business.

In the event of any adverse ruling in any intellectual property litigation, we could be required to pay substantial damages, cease the manufacturing, use and sale of infringing products, discontinue the use of certain processes or obtain a license from the third party claiming infringement with royalty payment obligations by us. Any litigation relating to the intellectual property rights of third parties, whether or not determined in our favor or settled by us, would be costly and may divert the efforts and attention of our management and technical personnel.

REGULATORY RISKS

The Failure to Complete Development of Our Medical Technology, Obtain Government Approvals, Including Required FDA Approvals, or to Comply With Ongoing Governmental Regulations Could Delay or Limit Introduction of Our Proposed Products and Result in Failure to Achieve Revenues or Maintain Our Ongoing Business

Our research and development activities, and the manufacture and marketing of our products, including our intended coronary artery bypass graft product, is subject to extensive regulation for safety, efficacy and quality by numerous government authorities in the United States and abroad. Before receiving clearance from the Food and Drug Administration ("FDA") to market our proposed graft, we will have to demonstrate that our grafts are safe and effective on the patient population. While we have done some preliminary animal trials and have seen acceptable results, there can be no assurance that acceptable results will be obtained in human trials. Clinical trials and manufacturing and marketing of medical devices are subject to the rigorous testing and approval processes of the FDA and equivalent foreign regulatory authorities. The Federal Food, Drug and Cosmetic Act and other federal, state and foreign statutes and regulations govern and influence the testing, manufacture, labeling, advertising, distribution and promotion of drugs and medical devices. As a result, clinical trials and regulatory approval of the coronary artery bypass graft can take a number of years or longer to accomplish and will require the expenditure of substantial financial, managerial and other resources.

In order to be commercially viable, we must successfully research, develop, obtain regulatory approval for, manufacture, introduce, market, and distribute our grafts. We must successfully meet a number of critical developmental milestones, including:

■ demonstrate benefit from the use of our grafts in various contexts such as coronary artery bypass surgery;
■ demonstrate through pre-clinical and clinical trials that our grafts are safe and effective; and
■ establish a viable good manufacturing process capable of potential scale-up.

The time-frame necessary to achieve these developmental milestones may be long and uncertain, and we may not successfully complete these milestones for any of our intended products in development.

Prior to conducting clinical trials, which are necessary to obtain approval by FDA to market a product, we must obtain FDA clearance. Even after obtaining clearance, the FDA can halt clinical trials at any time for safety reasons or for any violations, by us or by the clinical investigators, of the FDA's requirements for conducting clinical trials. If we are unable to receive clearance to conduct clinical trials, or if our clinical trials are halted by the FDA, we would not be able to achieve any revenue from such products since it is illegal to sell any medical device for human use without FDA approval.

Regulatory clearances or approvals, if granted, may include significant limitations on the indicated uses for which the product may be marketed. Prior to granting such clearances or approvals, the FDA and certain foreign regulatory authorities often impose numerous requirements with which medical device manufacturers must comply. FDA enforcement policy strictly prohibits the marketing of cleared or approved medical devices for uncleared or unapproved uses. In addition, product clearances or approvals could be withdrawn for failure to comply with regulatory standards or because of the occurrence of unforeseen problems following the initial marketing. We will be required to adhere to applicable FDA good manufacturing practice ("GMP") regulations and similar regulations in other countries, which include testing, control, and documentation requirements. Ongoing compliance with GMP and other applicable regulatory requirements will be monitored through periodic inspections by federal and state agencies, including FDA and CDHS, and by comparable agencies in other countries. Failure to comply with applicable regulatory requirements, including marketing products for unapproved uses, could result in, among other things, warning letters, fines, injunctions, civil penalties, recall or seizure of products, total or partial suspension of production, refusal of the government to grant pre-market clearance or pre-market approval for devices, withdrawal of clearances or approvals and criminal prosecution. Changes in existing regulations or adoption of new governmental regulations or policies could prevent or delay regulatory approval of our products.

There can be no assurance that we will be able to obtain FDA 510(k) clearance or PMA approval for our products under development or other necessary regulatory approvals or clearances on a timely basis or at all. Delays in receipt of or failure to receive U.S. or foreign clearances or approvals, the loss of previously obtained clearances or approvals, or failure to comply with existing or future regulatory requirements would likely have a material adverse effect on our business, financial condition, and results of operations.

We May Be Adversely Affected by Environmental Laws and Regulations

We are subject to a variety of laws, rules and regulations in the United States related to the use, storage, handling, discharge, and disposal of certain chemical materials such as isocyanates, dimethylacetamide, and glycols used in our research and manufacturing processes. Those regulations could potentially require us to acquire expensive equipment or to incur substantial other expenses to comply with them. If we incur such expenses, our product costs could significantly increase. Our failure to comply with present or future environmental laws, rules, and regulations could result in fines, suspension of production or cessation of operations. Although we are not aware of any claim involving violation of environmental or occupational safety and health laws and regulations, there can be no assurance that such a claim may not arise in the future.

OTHER RISKS

Our Future Success Depends on the Continued Service of Management, Engineering and Sales Personnel and Our Ability to Identify, Hire, and Retain Additional Personnel

Our success depends, to a significant extent, upon the efforts and abilities of Dr. Michael Szycher, our president and chief executive officer, and other members of senior management. The loss of the services of one or more of our senior management or other key employees could adversely affect our business. We do not maintain key person life insurance on any of our officers, employees, or consultants.

There is intense competition for qualified employees in the medical industry, particularly for highly skilled design, applications and engineers and sales people. We may not be able to continue to attract and retain technologists, managers, or other qualified personnel necessary for the development of our business or to replace qualified individuals who could leave us at any time in the future. Our anticipated growth is expected to place increased demands on our resources, and will likely require the addition of new management and engineering staff as well as the development of additional expertise by existing management employees. If we lose the services of or fail to recruit engineers or other technical and management personnel, our business could be harmed.

Periods of Rapid Growth and Expansion Could Place a Significant Strain on Our Resources, Including Our Employee Base

To manage our possible future growth effectively, we will be have to continue to improve our operational, financial and management systems. This process is complex and requires, among other things, that data from existing systems be made compatible with the upgraded systems. During transitional periods, we could experience delays in ordering materials, inventory tracking problems and other inefficiencies, which could cause delays in shipments of products to our customers.

In order to continue to grow, we will have to hire, train, motivate, and manage new and existing employees. In addition, our continued growth and the evolution of our business plan will require significant additional management, technical and administrative resources. We may not be able to effectively manage the growth and evolution of our current business.

The Anti-Takeover Provisions of Our Amended and Restated Certificate of Incorporation and of the Massachusetts Corporation Law May Delay, Defer, or Prevent a Change of Control

Our board of directors has the authority to issue up to 5,000,000 shares of preferred stock and to determine the price, rights, preferences and privileges and restrictions, including voting rights, of those shares without any further vote or action by our stockholders. The rights of the holders of our common stock will be subject to, and may be harmed by, the rights of the holders of any shares of preferred stock that may be issued in the future. The issuance of preferred stock may delay, defer or prevent a change in control because the terms of any issued preferred stock could potentially prevent or severely restrict us from the consummation of any merger, reorganization, sale of substantially all of our assets, liquidation or other extraordinary corporate transaction, without the approval of the holders of the then outstanding shares of preferred stock. In addition, the issuance of preferred stock could have a dilutive effect on our current stockholders.

Our stockholders must give substantial advance notice, prior to the relevant meeting, to nominate a candidate for director or present a proposal to our stockholders at a meeting. These notice requirements could inhibit a takeover by delaying stockholder action. In addition, our bylaws and Massachusetts law provide for staggered board members with each member elected for three years. In addition, directors may only be removed by stockholders for cause and with a vote of 80% of the stock.

Potential Healthcare Reform Legislation May Adversely Affect Our Business

Political, economic, and regulatory influences are subjecting the healthcare industry in the United States to fundamental change. Potential reforms proposed over the last several years have included mandated basic healthcare benefits, controls on healthcare spending through limitations on the growth of private health insurance premiums and Medicare and Medicaid spending, the creation of large insurance purchasing groups and other fundamental changes in the healthcare delivery system. In addition, some states in which we operate are also considering various healthcare reform proposals. We anticipate that federal and state governments will continue to review and assess alternative healthcare delivery systems and payment methodologies and that the public debate of these issues will likely continue in the future. Due to uncertainties regarding the ultimate features of reform initiatives and their enactment and implementation, we cannot predict which, if any, of such reform proposals will be adopted, when they may be adopted or what impact they may have on us and there can be no assurance that the adoption of reform proposals will not have a material adverse effect on our business, operating results or financial condition.

USE OF PROCEEDS

The Company will not receive any proceeds from sales by the selling security holders. However, we will receive funds from the selling security holders upon exercise of their warrants. The exercise price of the majority of these warrants is $1.50.

SELLING SECURITY HOLDERS

The following table and notes set forth, to the best of our knowledge, the name of each selling security holder, the nature of any position, office, or other material relationship which the selling security holder has had, within the past three years, with CardioTech or with any of our predecessors or affiliates, the amount of shares of CardioTech common stock that are beneficially owned by such security holder, the amount to be offered for the security holder's account and the amount to be owned by such security holder after completion of the offering.

Selling Security Holder	Number of Shares of CardioTech Common Stock Beneficially Owned Prior to the Offering	Number of Shares of CardioTech Common Stock Being Offered Hereby	Number of Shares of CardioTech Common Stock to be Owned Upon Completion of the Offering
Fechtor, Detwiler & Co. Inc.	100,000	100,000	0
Michael Adams	325,089	40,000	285,089
Michael Szycher	2,967,379	40,000	2,927,379
Alan Edwards*	80,047	40,000	40,047
John Mattern	40,000	40,000	0
Robert Detwiler	125,000	125,000	0
William Petty	70,000	25,000	45,000
Total	3,707,515	410,000	3,297,515

* Alan Edwards served as an officer of CardioTech, Ltd. prior to its sale in November of 2000.

Because the selling security holders may, under this prospectus, sell all or some portion of their CardioTech common stock, no estimate can be given as to the amount of CardioTech common stock that will be held by the selling security holders upon completion of the offering. In addition, the selling security holders identified above may have sold, transferred or otherwise disposed of all or a portion of their CardioTech common stock after the date on which they provided information regarding their share holdings.

PLAN OF DISTRIBUTION

Selling security holders may offer and sell, from time to time, the shares of our common stock covered by this prospectus. We refer to the common stock as the securities. The term selling security holders includes donees, pledgees, transferees or other successors-in-interest selling securities received after the date of this prospectus from a selling security holder as a gift, pledge, partnership distribution, or other non-sale related transfer. The selling security holders will act independently of us in making decisions with respect to the timing, manner, and size of each sale. Sales may be made on one or more exchanges or in the over-the-counter market or otherwise, at prices and under terms then prevailing or at prices related to the then current market price or in negotiated transactions. The selling security holders may sell their securities by one or more of, or a combination of, the following methods:

- purchases by a broker-dealer as principal and resale by the broker-dealer for its own account pursuant to this prospectus;
- ordinary brokerage transactions and transactions in which the broker solicits purchasers;
- block trades in which the broker-dealer so engaged will attempt to sell the securities as agent but may position and resell a portion of the block as principal to facilitate the transaction;
- an over-the-counter distribution in accordance with the rules of the Nasdaq National Market;
- in privately negotiated transactions; and
- in options transactions.

The shares of our common stock will be listed, and may be traded, on the American Stock Exchange under the symbol "CTE."

In addition, the selling security holders may sell any securities that qualify pursuant to Rule 144 under the Securities Act rather than pursuant to this prospectus.

To the extent required, we may amend or supplement this prospectus to describe a specific plan of distribution. In connection with distributions of the securities or otherwise, the selling security holders may enter into hedging transactions with broker-dealers or other financial institutions. In connection with those transactions, broker-dealers or other financial institutions may engage in short sales of shares of our common stock in the course of hedging the positions they assume with selling security holders. The selling security holders may also shares of our common stock short and redeliver the securities to close out their short positions. The selling security holders may also enter into option or other transactions with broker-dealers or other financial institutions that require the delivery to the broker-dealer or other financial institution of securities offered by this prospectus, which securities the broker-dealer or other financial institution may resell pursuant to this prospectus, as supplemented or amended to reflect the transaction. The selling security holders may also pledge securities to a broker-dealer or other financial institution, and, upon a default, the broker-dealer or other financial institution may affect sales of the pledged securities pursuant to this prospectus, as supplemented or amended to reflect the transaction.

In effecting sales, broker-dealers or agents engaged by the selling security holders may arrange for other broker-dealers to participate. Broker-dealers or agents may receive commissions, discounts, or concessions from the selling security holders in amounts to be negotiated immediately prior to the sale.

In offering the securities covered by this prospectus, the selling security holders and any broker-dealers who execute sales for the selling security holders may be treated as "underwriters" within the meaning of the Securities Act in connection with sales. Any profits realized by the selling security holders and the compensation of any broker-dealer may be treated as underwriting discounts and commissions. The selling security holders and any other person participating in a distribution will be subject to the Exchange Act. The Exchange Act rules include, without limitation, Regulation M, which may limit the timing of purchases and sales of any of the securities by the selling security holders and other participating persons. In addition, Regulation M may restrict the ability of any person engaged in the distribution of the securities to engage in market-making activities with respect to the particular security being distributed for a period of up to five business days prior to the commencement of the distribution. This may affect the marketability of the securities and the ability of any person or entity to engage in market-making activities with respect to the securities.

We will make copies of this prospectus available to the selling security holders for the purpose of satisfying the prospectus delivery requirements of the Securities Act, which may include delivery through the facilities of the

American Stock Exchange pursuant to Rule 153 under the Securities Act. The selling security holders may indemnify any broker-dealer that participates in transactions involving the sale of the securities against certain liabilities, including liabilities arising under the Securities Act.

At the time a particular offer of securities is made, if required, a prospectus supplement will be distributed that will set forth the number of securities being offered and the terms of the offering, including the name of any underwriter, dealer or agent, the purchase price paid by any underwriter, any discount, commission and other item constituting compensation, any discount, commission or concession allowed or re-allowed or paid to any dealer, and the proposed selling price to the public.

LEGAL MATTERS

The validity of the shares of CardioTech common stock being offered herein has been passed upon for CardioTech by Ellenoff Grossman & Schole LLP of New York, New York.

EXPERTS

CardioTech's consolidated balance sheet and the consolidated statements of income, retained earnings, and cash flows for each of the three years in the period ended March 31, 2003, incorporated in this prospectus by reference from CardioTech's Annual Report on Form 10-KSB for the year ended March 31, 2003 have been audited by Ernst and Young LLP, independent accountants, as stated in their report and have been so incorporated in reliance upon the report of such firm given upon their authority as experts in accounting and auditing.

The consolidated financial statements of CardioTech appearing in CardioTech's Annual Report on Form 10-KSB for the fiscal year ended March 31, 2002, incorporated by reference herein, have been audited by Arthur Anderson LLP, independent public accountants, as indicted in their report with respect thereto.

Effective August 31, 2002, Arthur Anderson LLP relinquished its license to practice in front of the SEC. Under these circumstances, Rule 437(a) under the Securities Act permits the registration statement to be filed without a written consent from Arthur Andersen LLP. The absence of such consent may limit your recovery on certain claims. In particular, and without limitation, you will not be able to assert claims against Arthur Andersen LLP under Section 11 of the Securities Act for any untrue statement of material fact contained in CardioTech's consolidated financial statements that appeared in its Annual Report on Form 10-KSB for the fiscal year ended March 31, 2002 or any omissions to state a material fact required to be stated therein. In addition, the ability of Arthur Andersen LLP to satisfy any claims (including claims arising from Arthur Andersen's provision of auditing and other services to CardioTech) may be limited as a practical matter due to recent events involving Arthur Andersen LLP.

WHERE YOU CAN FIND MORE INFORMATION

We file annual, quarterly, and special reports, proxy statements, and other information with the Securities and Exchange Commission. Our SEC filings are available over the Internet at the SEC's web site at http://www.sec.gov. You may also read and copy any document we file with the SEC at its public reference facilities: Public Reference Room, 450 Fifth Street, N.W., Room 1024, Washington, D.C. 20549.

You may also obtain copies of these documents at prescribed rates by writing to the Public Reference Section of the SEC at 450 Fifth Street, N.W., Room 1024, Washington, D.C. 20549. Please call 1-800-SEC-0330 for further information on the operations of the public reference facilities and copying charges.

We incorporate by reference into this prospectus the following documents filed by us with the SEC:

■ Quarterly Report on Form 10-QSB for the Quarter ended September 30, 2003
■ Quarterly Report on Form 10-QSB for the Quarter ended June 30, 2003
■ Annual Report on Form 10-KSB for the Fiscal Year ended March 31, 2003
■ Proxy Statement on Schedule 14A, filed on October 2, 2002

Any statement made in a document incorporated by reference or deemed incorporated herein by reference is deemed to be modified or superseded for purposes of this prospectus if a statement contained in this prospectus or in any other subsequently filed document which also is incorporated or deemed incorporated by reference herein modifies or supersedes that statement. Any such statement so modified or superseded shall not be deemed, except as so modified or superseded, to constitute a part of this prospectus. We also incorporate by reference into this prospectus all documents filed pursuant to Sections 13(a), 13(c), 14, or 15(d) of the Securities Exchange Act of 1934 after the date of this prospectus and prior to the termination of this offering.

The undersigned registrant hereby undertakes that, for purposes of determining any liability under the Securities Act of 1933, each filing of the registrant's annual report pursuant to section 13(a) or section 15(d) of the Securities Exchange Act of 1934 (and, where applicable, each filing of an employee benefit plan's annual report pursuant to section 15(d) of the Securities Exchange Act of 1934) that is incorporated by reference in the registration statement shall be deemed to be a new registration statement relating to the securities offered therein, and the offering of such securities at that time shall be deemed to be the initial bona fide offering thereof.

Statements made in this prospectus or in any document incorporated by reference in this prospectus as to the contents of any contract or other document referred to herein or therein are not necessarily complete, and in each instance reference is made to the copy of such contract or other document filed as an exhibit to the documents incorporated by reference, each such statement being qualified in all material respects by such reference.

This prospectus contains trademarks, trade names, service marks, and service names of CardioTech and other companies.

We will provide a copy of these filings and any exhibits specifically incorporated by reference in these filings at no cost upon written or oral request directed to us at the following address and telephone number: CardioTech International, Inc., 78-E Olympia Ave., Woburn, Massachusetts 01801 (781) 933-4772.

SPECIAL NOTE REGARDING FORWARD-LOOKING INFORMATION

This prospectus and the documents incorporated herein by reference contain "forward-looking statements" within the meaning of Section 27A of the Securities Act of 1933 and Section 21E of the Exchange Act. All statements other than statements of historical facts contained in this offering memorandum, including statements regarding our future financial position, business strategy, budgets, projected costs, and plans and objectives of management for future operations, are forward-looking statements. Although we believe that our expectations reflected in these forward-looking statements are based on reasonable assumptions, we cannot assure you that these expectations will prove to be correct. For detailed information regarding some of the risks, uncertainties, and assumptions that could cause actual results to differ materially from the expectations reflected in the forward-looking statements, please see "Risk Factors" beginning on page 4 of this prospectus.

We undertake no obligation to update or revise our forward-looking statements, whether as a result of new information, future events or otherwise. In light of these risks, uncertainties, and assumptions, you should be fully aware that the forward-looking events discussed in this offering memorandum might not occur.

PART II: INFORMATION NOT REQUIRED IN PROSPECTUS

EXPENSES OF ISSUANCE AND DISTRIBUTION

The following is an estimate of the expenses that we expect to incur in connection with this registration. We will pay all of these expenses, and the selling shareholders will not pay any of them.

SEC Registration fee		
Printing and engraving expenses	$1,000	*
Legal fees and expenses	$10,000	*
Accounting fees and expenses	$1,500	*
Miscellaneous		*
Total		*
Estimate, and subject to future contingencies.		

INDEMNIFICATION OF DIRECTORS AND OFFICERS

Indemnification under CardioTech's Certificate of Incorporation and By-laws and Massachusetts Law. CardioTech's certificate of incorporation provides for the indemnification of directors or officers, in accordance with the by-laws, to the fullest extent permitted by the Massachusetts General Corporation Law. CardioTech's bylaws also provides that CardioTech shall indemnify and hold harmless, to the fullest extent permitted by law, any person made or threatened to be made a party to any legal action by reason of the fact that such person is or was a director, officer, employee or other corporate agent of CardioTech or any subsidiary or constituent corporation or served any other enterprise at the request of CardioTech, against expenses, judgments, fines and amounts paid in settlement actually and reasonably incurred by such person in connection with such action. Massachusetts General Corporation Law provides for the indemnification of directors and officers under certain conditions.

CardioTech Directors and Officers Insurance. The directors and officers of CardioTech are insured under a policy of directors' and officers' liability insurance.

EXHIBITS AND FINANCIAL STATEMENT SCHEDULES

The following is a list of Exhibits included as part of this Registration Statement. CardioTech agrees to furnish supplementally a copy of any omitted exhibit or schedule to the Commission upon request.

Exhibit No.	Exhibit Title
2.1	Agreement and Plan of Merger dated as of October 25, 2002 among CardioTech International, Inc. and Gish Biomedical, Inc. and Gish Acquisition Corp., incorporated by reference to the Joint Proxy Statement/Prospectus filed by CardioTech International, Inc. as part of its Registration Statement on Form S-4 filed on December 23, 2002.
2.2	Amendment No. 1 to the Agreement and Plan of Merger dated as of January 9, 2003 among CardioTech International, Inc., Gish Biomedical, Inc. and Gish Acquisition Corp., incorporated by reference to Exhibit 2.2 to Amendment No. 1 to CardioTech's Registration Statement on Form S-4, filed on January 16, 2003.

3.1	Restated Articles of Organization of CardioTech International, Inc., incorporated by reference to Exhibit 3.1 to Amendment No. 1 to CardioTech International, Inc.'s Registration Statement on Form 10-12G, filed on May 10, 1996.
3.2	Amended and Restated By-Laws of CardioTech International, Inc., incorporated by reference to Exhibit 3.2 to CardioTech International, Inc.'s Registration Statement on Form 10-12G, filed on March 20, 1996.
5.1	** Opinion of Ellenoff Grossman & Schole, LLP for CardioTech International, Inc., as to the legality of the securities being registered.
10.2	Tax Matters Agreement between Poly Medica Industries, Inc. and CardioTech International, Inc., dated as of May 13, 1996, incorporated by reference to Amendment No. 1 to CardioTech International, Inc.'s Registration Statement on Form 10-12G, filed on May 10, 1996.
10.3	Amended and Restated License Agreement between PMI and CardioTech International, Inc., dated as of May 9, 1996, incorporated by reference to Exhibit 10.4 to Amendment No. 1 to CardioTech International, Inc.'s Registration Statement on Form 10-12G, filed on May 10, 1996.
10.4	1996 Employee, Director and Consultant Stock Option Plan, incorporated by reference to Exhibit 10.4 to CardioTech International, Inc.'s Annual Report on Form 10-K for the year ended March 31, 1998, filed on June 29, 1998.
10.5	Employment Agreement, dated as of March 26, 1998, between Michael Szycher, Ph.D., and CardioTech, incorporated by reference to Exhibit 10.5 to CardioTech International, Inc.'s Annual Report on Form 10-K for the year ended March 31, 1998, filed on June 29, 1998.
10.6	Development, Supply and License Agreement between PMI and Bard Access Systems, dated November 11, 1992, incorporated by reference to Exhibit 10.10 to CardioTech International, Inc.'s Registration Statement on Form 10-12G, filed on March 20, 1996.
10.7	Commercial Lease between CardioTech International, Inc. and Cummings Properties Management, Inc., dated June 26, 1998, incorporated by reference to Exhibit 10.11 to CardioTech International, Inc.'s Annual Report on Form 10-K for the year ended March 31, 1998, filed on June 29, 1998.
10.8	Form of Unit Purchase Agreement, incorporated by reference to Exhibit 99.1 to CardioTech International, Inc.'s Registration Statement on Form S-3 filed on February 12, 1999.

10.9	Form of Warrant to purchase shares of CardioTech International, Inc. common stock, incorporated by reference to Exhibit 99.2 to CardioTech International, Inc.'s Registration Statement on Form S-3, filed on February 12, 1999.
23.1	** Consent of Ernst & Young LLP, Independent Auditors.

**Filed herewith*

ITEM 22. UNDERTAKINGS.

(a) The undersigned registrant hereby undertakes:

(1) To file, during any period in which offers or sales are being made, a post-effective amendment to this registration statement

(i) To include any prospectus required by Section 10(a)(3) of the Securities Act of 1933;

(ii) To reflect in the prospectus any facts or events arising after the effective date of the registration statement (or the most recent post-effective amendment thereof) which, individually or in the aggregate, represent a fundamental change in the information set forth in the registration statement. Notwithstanding the foregoing, any increase or decrease in volume of securities offered (if the total dollar value of securities offered would not exceed that which was registered) and any deviation from the low or high end of the estimated maximum offering range may be reflected in the form of prospectus filed with the Commission pursuant to Rule 424(b) if, in the aggregate, the changes in volume and price represent no more than a 20 percent change in the maximum aggregate offering price set forth in the "Calculation of Registration Fee" table in the effective registration statement;

(iii) To include any material information with respect to the plan of distribution not previously disclosed in the registration statement or any material change to such information in the registration statement.

(2) That, for the purpose of determining any liability under the Securities Act of 1933, each such post-effective amendment shall be deemed to be a new registration statement relating to the securities offered therein, and the offering of such securities at that time shall be deemed to be the initial bona fide offering thereof.

(3) To remove from registration by means of a post-effective amendment any of the securities being registered which remain unsold at the termination of the offering.

(4) That prior to any public reoffering of the securities registered hereunder through use of a prospectus which is a part of this registration statement, by any person or party who is deemed to be an underwriter within the meaning of Rule 145(c), such reoffering prospectus will contain the information called for by the applicable registration form with respect to reofferings by persons who may be deemed underwriters, in addition to the information called for by the other items of the applicable form.

(5) That every prospectus (i) that is filed pursuant to paragraph (4) immediately preceding, or (ii) that purports to meet the requirements of Section 10(a)(3) of the Securities Act of 1933 and is used in connection with an offering of securities subject to Rule 415, will be filed as a part of an amendment to the registration statement and will not be used until such amendment is effective, and that, for purposes of determining any liability under the Securities Act of 1933, each such post-effective amendment shall be deemed to be a new registration statement relating to the securities offered therein, and the offering of such securities at that time shall be deemed to be the initial bona fide offering thereof.

(6) That, for purposes of determining any liability under the Securities Act of 1933, each filing of the registrant's annual report pursuant to Section 13(a) or 15(d) of the Securities Exchange Act of 1934 (and, where applicable, each filing of an employee benefit plan's annual report pursuant to Section 15(d) of the Securities Exchange Act of 1934) that is incorporated by reference in the registration statement shall be deemed to be a new registration statement relating to the securities offered therein, and the offering of such securities at that time shall be deemed to be the initial bona fide offering thereof.

(7) To respond to requests for information that is incorporated by reference into the joint proxy statement/prospectus pursuant to Item 4, 10(b), 11 or 13 of this form, within one business day of receipt of such request, and to send the incorporated documents by first class mail or other equally prompt means. This includes information contained in documents filed subsequent to the effective date of the registration statement through the date of responding to the request.

(8) To supply by means of a post-effective amendment all information concerning a transaction, and the company being acquired involved therein, that was not the subject of and included in the registration statement when it became effective.

(b) Insofar as indemnification for liabilities arising under the Securities Act of 1933 may be permitted to directors, officers and controlling persons of the registrant pursuant to the foregoing provisions, or otherwise, the registrant has been advised that in the opinion of the Securities and Exchange Commission such indemnification is against public policy as expressed in the Act and is, therefore, unenforceable. In the event that a claim for indemnification against such liabilities (other than the payment by the registrant of expenses incurred or paid by a director, officer or controlling person of the registrant in the successful defense of any action, suit or proceeding) is asserted by such director, officer or controlling person in connection with the securities being registered, the registrant will, unless in the opinion of its counsel the matter has been settled by controlling precedent, submit to a court of appropriate jurisdiction the question whether such indemnification by it is against public policy as expressed in the Act and will be governed by the final adjudication of such issue.

SIGNATURES

Pursuant to the requirements of the Securities Act of 1933, as amended, the registrant certifies that it has reasonable grounds to believe that it meets all of the requirements for filing on Form S-3 and has duly caused this registration statement to be signed on its behalf by the undersigned, thereunto duly authorized, in the City of Woburn, State of Massachusetts, on November 12, 2003.

CARDIOTECH INTERNATIONAL, INC.

By: /s/ MICHAEL SZYCHER

Michael Szycher, Ph.D.

Chairman and Chief Executive Officer

POWER OF ATTORNEY

KNOW ALL PERSONS BY THESE PRESENTS, that each of the undersigned hereby constitutes and appoints Michael Szycher, Ph.D. as his true and lawful attorney-in-fact and agent, with full power of substitution and resubstitution, for him and on his behalf to sign, execute and file this registration statement and any or all amendments (including, without limitation, post-effective amendments) to this registration statement, and to file the same, with all exhibits thereto and any and all documents required to be filed with respect therewith, with the Securities and Exchange Commission or any regulatory authority, granting unto such attorney-in-fact and agent full power and authority to do and perform each and every act and thing requisite and necessary to be done in connection therewith and about the premises in order to effectuate the same as fully to all intents and purposes as he might or could do if personally present, hereby ratifying and confirming all that such attorney-in-fact and agent, or his substitute or substitutes, may lawfully do or cause to be done.

Pursuant to the requirements of the Securities Act of 1933, as amended, this registration statement has been signed below by the following persons on behalf of the Registrant and in the capacities and on the dates indicated:

Signature	Title	Date
/s/ MICHAEL SZYCHER Michael Szycher	Chairman, Chief Executive Officer and Director (Principal Executive Officer)	November 25, 2003
/s/ THOMAS LOVETT Thomas Lovett	VP of Finance (Principal Financial Officer)	November 25, 2003
/s/ MICHAEL ADAMS Michael Adams	Director	November 25, 2003
/s/ ANTHONY J. ARMINI Anthony J. Armini	Director	November 25, 2003
/s/ MICHAEL BARRETTI Michael Barretti	Director	November 25, 2003

EXHIBIT INDEX

2.1	Agreement and Plan of Merger dated as of October 25, 2002 among CardioTech International, Inc. and Gish Biomedical, Inc. and Gish Acquisition Corp., incorporated by reference to the Joint Proxy Statement/Prospectus filed by CardioTech International, Inc. as part of its Registration Statement on Form S-4 filed on December 23, 2002.
2.2	Amendment No. 1 to the Agreement and Plan of Merger dated as of January 9, 2003 among CardioTech International, Inc., Gish Biomedical, Inc. and Gish Acquisition Corp., incorporated by reference to Exhibit 2.2 to Amendment No. 1 to CardioTech's Registration Statement on Form S-4, filed on January 16, 2003.

3.1	Restated Articles of Organization of CardioTech International, Inc., incorporated by reference to Exhibit 3.1 to Amendment No. 1 to CardioTech International, Inc.'s Registration Statement on Form 10-12G, filed on May 10, 1996.
3.2	Amended and Restated By-Laws of CardioTech International, Inc., incorporated by reference to Exhibit 3.2 to CardioTech International, Inc.'s Registration Statement on Form 10-12G, filed on March 20, 1996.
5.1	** Opinion of Ellenoff Grossman & Schole, LLP for CardioTech International, Inc., as to the legality of the securities being registered.
10.2	Tax Matters Agreement between Poly Medica Industries, Inc. and CardioTech International, Inc., dated as of May 13, 1996, incorporated by reference to Amendment No. 1 to CardioTech International, Inc.'s Registration Statement on Form 10-12G, filed on May 10, 1996.
10.3	Amended and Restated License Agreement between PMI and CardioTech International, Inc., dated as of May 9, 1996, incorporated by reference to Exhibit 10.4 to Amendment No. 1 to CardioTech International, Inc.'s Registration Statement on Form 10-12G, filed on May 10, 1996.
10.4	1996 Employee, Director and Consultant Stock Option Plan, incorporated by reference to Exhibit 10.4 to CardioTech International, Inc.'s Annual Report on Form 10-K for the year ended March 31, 1998, filed on June 29, 1998.
10.5	Employment Agreement, dated as of March 26, 1998, between Michael Szycher, Ph.D., and CardioTech, incorporated by reference to Exhibit 10.5 to CardioTech International, Inc.'s Annual Report on Form 10-K for the year ended March 31, 1998, filed on June 29, 1998.
10.6	Development, Supply and License Agreement between PMI and Bard Access Systems, dated November 11, 1992, incorporated by reference to Exhibit 10.10 to CardioTech International, Inc.'s Registration Statement on Form 10-12G, filed on March 20, 1996.
10.7	Commercial Lease between CardioTech International, Inc. and Cummings Properties Management, Inc., dated June 26, 1998, incorporated by reference to Exhibit 10.11 to CardioTech International, Inc.'s Annual Report on Form 10-K for the year ended March 31, 1998, filed on June 29, 1998.
10.8	Form of Unit Purchase Agreement, incorporated by reference to Exhibit 99.1 to CardioTech International, Inc.'s Registration Statement on Form S-3 filed on February 12, 1999.

Exhibit 5.1

November 25, 2003

Cardiotech International, Inc.
78-E Olympia Avenue
Woburn, MA 01801

Ladies and Gentlemen:

We have acted as counsel for Cardiotech International, Inc., a Massachusetts corporation (the "Company"), in connection with the preparation of its Registration Statement on Form S-3 (the "Registration Statement") to which this opinion is an exhibit, to be filed with the Securities and Exchange Commission on or about November 25, 2003.

It is our opinion that the shares of Common Stock, par value $0.01 per share, being registered, when sold in the manner and for the consideration contemplated by the Registration Statement, will be validly issued, fully paid and non-assessable.

We consent to the filing of this opinion as an Exhibit to the Registration Statement and to the reference to our firm under the heading "Legal Matters."

Very truly yours,

Ellenoff Grossman & Schole LLP

By:/s/Ellenoff Grossman & Schole LLP

Consent of Independent Auditors

We consent to the reference to our firm under the caption "Experts" in the Registration Statement Form S-3 for the registration of 410,000 shares of its common stock and to the incorporation by reference therein of our report dated May 16, 2003 with respect to the financial statements of CardioTech International, Inc. and our report dated May 23, 2003 with respect to the financial statements of Gish Biomedical, Inc. for the year ended March 31, 2003 included in CardioTech International, Inc.'s Annual Report on Form 10-KSB.

/s/Ernst & Young, LLP
Boston, Massachusetts
November 26, 2003

End of prospectus

References

1. Wennberg, K., et al. Reconceptualizing entrepreneurial exit: Divergent exit routes and their drivers. *J Business Venturing,* 2009, doi :10.1016/j. jbusvent.2009.01.001.
2. http://www.investopedia.com/terms/c/capitalization.asp
3. http://en.wikipedia.org/wiki/Strategic_alliance
4. Andrade, G., Mitchell, M., & Stafford, E. New evidence and perspectives on mergers. *J Econ Perspectives,* 15(2), 103–120, 2001.
5. http://en.wikipedia.org/wiki/Mergers_and_acquisitions
6. http://en.wikipedia.org/wiki/Mergers_and_acquisitions
7. America Online and Time Warner Will Merge to Create World's First Internet-Age Media and Communications Company. Time Warner corporate homepage.
8. Houghton, J.R. Corning cultivates joint ventures that endure. *Planning Review,* 18(5), Sept/Oct 15–17, 1990.
9. The Free Dictionary. http://financial-dictionary.thefreedictionary.com/ Accretive+Acquisition.

Index

A

ABCs of entrepreneurial investments, 134
Academic research, 37–38
Accommodating negotiation style, 166
Accountants, 142
Accredited investors, 126
Accretive acquisition, 357, 361
Acquisition, 350–351, *See also* Mergers and
      acquisitions
  accretive, 357, 361
  acquisition criteria matrix, 353–354
  case study, 352–362
  SWOT analysis, 357–361
Adoption cycle, 188–189
Affordable Healthcare Act, 124
Agenda for negotiation, 162–163
America Invents Act (AIA), 215–217
American Research and Development
      Corporation (ARDC), 127
America Online (AOL)-Time Warner
      merger, 351–352
Angel investors, 46, 126
  female entrepreneurs and, 47
  series "A" round, 258–259, 338
Arbitration, 154
Asset-based valuation approach, 263
AT&T, 36
Auction tactic in negotiation, 167
Authority, 68, 69
Autocratic leadership style, 63
Available market, 185
Avoiding negotiation style, 166

B

Bargaining, 154
Bayh-Dole Act of 1980, 37
Benefits program, 123–124, 125*f*
Best alternative to a negotiated agreement
      (BATNA), 161–162
Biotechnology patents, 218–221
Blake-Mouton Managerial Grid, 83–84
Blank, Arthur, 60
Board of advisors (BOA), 66–67
Board of directors (BOD), 65–66, 137
Body language, 172
Bogey tactic in negotiation, 168
Bootstrapping, 257
Boston Consulting Group (BCG) growth-
      share matrix, 250–252
Breakthrough innovations, 104–105
Brinksmanship, 167–168
Brown, Sarah, 47–50
  interview with, 51–54
Budapest Treaty, 221
Business plan, 273–274
  components, 276–277, 278–280*t*
  example, 285–334
  implementation potential, 284
  irresistible offer, 280
  milestones, 284
  need for, 274–275, 276*f*
  risk mitigation, 277
  types, 274*f*
Business presentations, 128–133,
      174–175

C

Capitalization, 337–339, *See also* Financing
    going public, 339–343
Capitalization rate, 262
Capitalized returns method, 260*f*, 261–263
Carbon footprint, 50
Cardiotech, 352–362
    prospectus, 362–393
Carnegie, Andrew, 60
Cash flow, 101–102
CBS, 36
Change
    overcoming inertia and defenses,
        147–150
    resistance to, 75, 76*t*
    unfreeze, change, refreeze model,
        76–77
Change agents, 27, 75
Chemical sciences patents, 220
Chicken or egg cycle, 170–171
Chicken tactic in negotiation, 168
Chief executive officer (CEO), 121
Chief financial officer (CFO), 122
Chief operations officer (COO), 122
Christopher Columbus syndrome, 276
Churchill and Lewis's growth model,
    94–96
Cincinnati Bell, 36
Closing the deal, 175–177
Coercive power, 69
Cohen, Ben, 86
Collaborating negotiation style, 166
Commercialization, 103
Commoditization, 193, 194*f*
Company name, 117–118
Compensation, 123–124
    indemnification of directors and officers,
        386
    JOBS Act and disclosure requirements,
        235–236
    sales force, 198–199
Competing negotiation style, 166
Competitive intelligence (CI), 244–246
Complex sales, 193
Compromising negotiation style, 166

Conflict management, 82–84, 173–174
Conoco, 36
Conservatives, 188
Continuation patent applications, 214
Controller, 123
Convergys, 36
Cooperative venture, 344
Copyright, 204, 205*t*
Core competencies, 242
Core purpose, 242
Core values, 242
Corporate culture and innovation, 26–28,
        103–104, 106–107
Corporate entrepreneurship, *See*
    Intrapreneurship
Corporate executive duties and
    responsibilities, 121–123
Corporate governance, 65–66
Corporate legal structure, 118–121
Corporate life cycles, 90–91
    early adopters, 111*f*, 113–114
    growth models, 93–96
Corporate secretary, 122
Cost approach to valuation, 263
Cost leadership strategy, 239
Covenants, 348
Crises, 93–94
Critical mass, 183
Crossing the chasm, 189–190
Crowdfunding, 229–233
Culture of innovation, 26–28, 103–104,
    106–107
Customer critical mass, 183

D

Data, information, knowledge, and
    know-how (DIKK), 28–30
Deadlines tactic in negotiation, 168
Deadlocks, 170
Debt financing vs. equity financing, 130*t*
Defense in depth, 168
Dell, 3
Dell, Michael, 60
Demands, 184–185
Democratic leadership style, 63

Dependent patent claims, 210
Differentiation, 194, 195, 239
Digital Equipment Corporation (DEC), 127
DIKK, 28–30
Discounted cash flow rate of return
   (DCFROR), 259
Discounted future returns method, 261
Disruptive innovation, 104–105
Disruptive technology, 100*t*, 105
Distributive negotiation, 156–157
Diversity, 5
Divestiture, 30, 34
Divisional patent applications, 214
Dodd-Frank Act, 228, 235
Dominant coalition, 72, 74–75
Due diligence, 127
DuPont, 36

E

Early adopters, 111*f*, 113–114, 188
Early majority, 114
Early stage VC financing, 266
Economic rate of return (ERR), 259
Edison, Thomas Alva, 61, 282
Elevator pitch, 131–132
Emotional support, 52
Empire builder entrepreneur, 86
Employee benefits, 123–124, 125*f*
Empowerment, 148
Enthusiasts, 114
Entrepreneurial hero, 84–85
Entrepreneurial rewards, 84–85
Entrepreneurs, 1, 57
  as chief negotiators, 157
  common characteristics, 3*t*
  the crucial question, 12*f*
  crucial skills for, 57, 58*f*
  differences between intrapreneurs, 20*t*
  diversity, 5
  female, *See* Women and
    entrepreneurship
  frequently asked questions, 6–8
  leaders vs. managers, 59–60
  male and female entrepreneur
    characteristics, 42*t*
  motivations, 4
  rules for, 14*f*
  success characteristics, 9*t*
  types, 85–86
Entrepreneurship, 1–2, *See also* New
    venture creation
  failure statistics, 2, *See also* Failure
  global activity, 4
  ten commandments of, 15*f*
Entrepreneurship, corporate, *See*
    Intrapreneurship
Entrepreneurship, women and, *See* Women
    and entrepreneurship
Entrepreneur spin, 270
Equity carve-out (ECO), 30, 34–36
Equity financing vs. debt financing, 130*t*,
    See also Financing
Ethos, 129
European Patent Office (EPO), 214
Executive team, 139–141
  duties and responsibilities, 121–123
  managing managers, 143–145
  too many chiefs, 141–142
Exits (liquidity events), 264–265
Exit strategy, 12, 13*t*, 335, 351–352, *See also*
    Mergers and acquisitions
  harvesting, 335
  reasons for exiting, 336–337
  sample prospectus, 362–393
  seller/buyer frequently asked questions,
    353*t*
  start at the end, 336
Expenses, estimated startup costs, 133
Expert power, 69
External corporate venturing, 30–31

F

Failure
  "fail fast," 145–146
  fear of, 147–148
  management incompetence and, 5, 10*t*
  merger and acquisition, 351–352
  reasons for new ventures, 137
  small business survival ratios, 2
Fair market value, 260

Fair value standard of value, 262
Family support, 52
Female-owned businesses, 41, *See also* Women and entrepreneurship
  Green Alliance example, 47–50
  interview, 51–54
Financial management plan, 277
Financing, *See also* Angel investors; Valuation; Venture capital
  ABCs of entrepreneurial investments, 134
  basis for venture investment, 215
  capitalization, 337–339
  crowdfunding, 229–231
  differences between debt and equity, 130t
  females and small business loans, 45–46
  friends and family, 125–126
  fundraising importance of patents, 224–225
  going public, 339–343
  JOBS Act, 227–238
  persuasive presentations, 128–133
  principles of raising capital, 127–128
  types, 129f
Fiorina, Carly, 59
First draft of document, 163
First movers, 93
First public disclosure, 224
Flinch tactic in negotiation, 168
Focus strategy, 239
Follower entry strategy, 93
Followership, 77–80
Forbes, Malcolm, 86
Force field analysis (FFA), 77
For-profit corporations, 120
Founder organizational persona, 72–74
Founder's syndrome, 256
Founding team, 64–65, 137–138
Friends and family financing, 125–126
Fry, Art, 21, 22

G

Gates, Bill, 60, 86
Generalists, 93

Genetic engineering patents, 220
Gish Biomedical, 354–362
Global Fortune 500 company turnover, 91, 92t
Goals, 241
  executive team and, 140
Going public, 339–343
Good guy/bad guy tactic in negotiation, 168
Governance, 65–66
Great market squeeze, 193, 194f
Green Alliance, 47–50
  interview with owner, 51–54
Green business enterprises, 50
Greiner's model of organizational growth, 93–94
Growth-share matrix, 250–252

H

Hall of Shame, 282–283
Hard bargainers, 154, 156
Harvesting, 335, *See also* Exit strategy
  sample prospectus, 362–393
  seller/buyer frequently asked questions, 353t
Hewlett Packard, 3
Highball/lowball tactic in negotiation, 169
Hurdle rate, 268

I

Iacocca, Lee, 59
Idea mining, 89
Impasse situation, 170
Income approach to valuation, 260–263
Incorporation, 118
Incremental innovation, 104
Indemnification of directors and officers, 386
Independent patent claims, 210
Industrial design right, 204
Inertia and defenses, 147–150
Infinity Broadcasting, 36
Influence, 68
Information, 244

Information sharing, 176
Initial public offerings (IPOs), 266, 339
  equity carve-outs, 35–36
  JOBS Act and confidential filing process, 236–237
  JOBS Act and emerging growth companies, 234–235
Innovation, 103, *See also* Intellectual property
  classical adoption cycle, 188–189
  corporate culture and, 26–28, 103–104, 106–107, *See also* Intrapreneurship
  early adopters, 111*f*, 113–114
  exaggeration of technology, 270
  idea mining, 89
  invention, 103
  naysayers, 282–283
  new product life cycles, 96–99
  new venture opportunities, 89–90, *See also* New venture creation
  organizational culture and, 26–28
  overcoming inertia and defenses, 147–150
  process, 31*f*
  promoting, 183, *See also* Marketing
  transforming an idea into know-how, 28–30
  types, 103–106
  "why" questions of investors, 282
Innovator entrepreneur, 86
Integrative negotiation, 157
Intellectual property (IP), 203–206, *See also* Patents
  academic research and, 37
  advice for entrepreneurs, 221
  America Invents Act, 215–217
  basis for venture investment, 215
  benefits of, 207
  biotechnology, 218–221
  first public disclosure, 224
  risk information statement, 372–373
  trade secrets, 205*t*, 206, 225
Intelligence, 244, *See also* Marketing intelligence
Internal corporate venturing, 30–31
Internal rate of return (IRR), 259, 267, 268

International patent organizations, 214
Intrapreneurship, 1, 17, 20, 25
  academic research, 37–38
  change agents, 27
  definitions, 17–18
  differences between intrapreneurs and entrepreneurs, 20*t*
  encouraging, 23, 24*t*, 28, 29*t*
  equity carve-outs, 30, 34–36
  innovation culture, 26–28
  internal and external business environments, 25–26
  intrapreneur characteristics, 25*t*
  "Intraprise Manifesto," 18
  rules for Skunk Works, 18–20
  spin-outs and spin-offs, 30–34
  successful product example, 21–23
  Ten Commandments of, 38–39
  transforming an idea into know-how, 28–30
Invention, 103, 204, *See also* Innovation; Intellectual property; Patents
  America Invents Act, 215–217
  conception and inventorship, 222–223
  inventors in your company, 222
  patentable, 206–207
  reduction to practice, 222
Investor communications restrictions, 237–238
Invitational leadership, 63

J

J.H. Whitney & Company, 127
Jobs, Steve, 60, 183
JOBS Act, 227–238
  confidential filing process, 236–237
  emerging growth company IPOs, 234–235
  funding portals, 232–233
  information access and disclosure requirements, 229–230
  intermediary requirements, 231–232
  investor communications restrictions, 237–238
  issuer requirements, 231

Johnson, Kelly, rules for Skunk Works, 18–20
Johnson & Johnson, 145–146, 243–244
Joint ownership, 344
Joint venture, 344

K

Kellar, Andrew, 49

L

Laggards, 114, 188–189
Laissez-faire leadership style, 63
Last-minute demands, 177
Late majority, 114
Later stage VC financing, 266
Lawyers, 142–143
Leadership, 59
    authority, power, and influence, 68–69
    conflict management, 82–84
    defining, 59
    executive duties and responsibilities, 121–123
    executive team, 139–141
    followership, 77–80
    founder organizational persona, 72–74
    importance for startup businesses, 61–62
    leaders vs. managers, 59–60
    roles, 59*t*
    situational, 62–63
    skills for senior managers, 144
    styles/types, 63, 83
    Ten Commandments of, 86–87
    too many chiefs, 141–142
Leadership Grid, 83–84
Leahy-Smith America Invents Act (AIA), 215–217
Legal structures, 118–121
Legal team, 142–143
Legal terminology of corporations, 119*t*
Legitimate power, 68–69
Lewin, Kurt, 76–77
Liability risks information statement, 372
License agreements, 223
Life cycle of a venture, *See* Corporate life cycles

Lifestyle entrepreneur, 86
Liquidity, 101–102
Liquidity events, 264–265
Logos, 129

M

Management, 143–145
Management incompetence, 5, 10*t*
Management team, 280–281
Managerial Grid, 83–84
Managers vs. leaders, 59–60
Marcus, Bernie, 60
Market approach to valuation, 263
Market categories, 185
Marketing, 181, 183
    analyzing the market, 185
    costs and functions, 187*t*
    crossing the chasm, 189–190
    definition, 187
    differentiation, 194, 195
    mass and target, 250
    needs, wants, and demands, 184–185
    pioneering startups, 188–190
    positioning, 195–196
    positioning statements, 196–197
    promoting innovative products, 183
    pull and push strategies, 181–182
    strategic marketing goals, 250–252
    strategic orientations, 186*t*
    tensions with sales, 192–193
    unmet needs and wants, 186
Marketing intelligence (MI), 183, 246–247
Marketing myopia, 186, 187*t*
Marketing plan, 277
Market research, 183
Market risks information statement, 370–371
Market segmentation, 194–195, 196*t*
Market share, 248
Mass marketing, 250
Mediation, 154
Medical sciences patents, 220
Mergers and acquisitions (M&A), 345, 350–351
    accretive acquisition, 357, 361
    acquisition case study, 352–362

acquisition criteria matrix, 353–354
AOL-Time Warner, 351–352
best practices, 347–349
opportunities, 345
success/failure, 351–352
SWOT analysis, 357–361
waves, 351*f*
Mezzanine, 266
Microbiological sciences patents, 220, 221
Microsoft, 3
Milestones, 284
Minimum attractive rate of return, 268
Mission, 107–108, 242
    Green Alliance, 50
    Johnson & Johnson, 145–146, 243–244
MIT, 37–38
Monsanto, 36
Moore, Gordon, 86
Motivations of entrepreneurs, 4
    female entrepreneurs, 43
Mythos, 129

N

Naming the product, 118
Naming your company, 117–118
Needs, 184, 186
Negotiation, 153–154
    agenda, 162–163
    aligning interests, 161
    bargaining, 154
    body language and assessing progress, 172
    closing the deal, 175–177
    conflict management, 173–174
    distributive strategy, 156–157
    document first draft, 163
    expectations and, 153, 159
    founder as chief negotiator, 157
    founder's negotiation wheel, 155*f*
    impasse, stalemate, and deadlock situations, 170
    integrative strategy, 157
    knowing your counterpart, 159–160
    laws of power negotiation, 169–170
    measuring success, 164
    mediation and arbitration vs., 154

    negotiating cycle, 158*f*
    physical environment, 171–172
    power and, 154, 157–158
    power presentations, 174–175
    soft, hard, and principled, 154–156
    startup fears, 158
    strategies and guidelines, 164–165
    styles, 166
    success planning terminology, 161–162, 163*f*
    tactics, 166–169
    Ten Amendments of, 177–178
    time, information, and reward power, 173
    time-related pressures, 170–171
    valuation, 259
    your first negotiation, 160–161
Net income capitalization rate, 262
New product life cycles, 96–99
New venture creation, *See also* Startup business ventures
    bridging the Valley of Death, 114*f*, 115
    early adopters, 111*f*, 113–114
    idea mining, 89
    legal structures, 118–121
    life cycles, 90–99, *See also* Corporate life cycles
    naming your company, 117–118
    opportunities, 89–90, *See also* Innovation
    pioneering startups, 188–190
    resource issues, 99–102
    statistics regarding, 89
    strategic planning, 107–108
    types of innovation, 103–106
    value chain analysis, 108–109
    value proposition, 109–111
Nibble tactic in negotiation, 169
Non-disclosure agreement (NDA), 217
Not-for-profit corporations, 120

O

Operational planning, 108
Operations and management plan, 277
Organizational defenses, 149–150
Organizational growth models, 93–96

Organizational inertia, 148
Organizational persona, 72–74
Organizational politics, 69–73
Organizational resistance to change, 75,
    76t
Organization chart, 70f
Outcomes focus, 141

P

Parallel organization, 70, 71f
Pareto's law (80/20 rule), 81–82, 175–177
Patent Cooperation Treaty (PCT), 214
Patent Pending, 212
Patents, 203–204, 205t, 207–208
    advice for entrepreneurs, 221
    America Invents Act, 215–217
    benefits of, 207
    biotechnology, 218–221
    common elements, 208
    conception and inventorship, 222–223
    divisional and continuation applications,
        214
    first public disclosure, 224
    first to file, 215–216
    fundraising importance of, 224–225
    international organizations, 214
    licensing, 223
    patentable inventions, 206–207
    patent claims, 208–211
    provisional, 212–213
    similarity to real estate deeds, 212t
Paternalistic leadership style, 63
Pathos, 129
Persuasive business presentations,
    128–133
Pharmaceutical sciences patents, 220
Pharmacia, 36
Pinchot, Elizabeth, 17–18
Pinchot, Gifford, III, 17–18
Pioneer strategy, 190
Pizza analogy, 248–249
Planning, *See* Business plan; Strategic
    planning
Plant and animal sciences patents, 220
Politics, 69–73

Polymedica Industries, Inc., business plan,
    285–334
Positioning, 195–196
Positioning statements, 196–197
Post-it Notes®, 21–23, 207
Post-Money valuation, 134–135
Potential market, 185
Power, 68–69
    authority vs., 69
    dominant coalition, 72, 74–75
    negotiation and, 154, 157–158
Power negotiations, 153, 169–170, *See also*
        Negotiation
    time, information, and reward power, 173
Pragmatists, 114, 188–189
Pre-Money valuation, 134–135
Presentations, 128–133, 174–175
President, 121
Pricing, 190–192
    don't undersell, 52
    sales-marketing tensions, 192–193
    value vs. price, 256
Primary activities, 108–109
Principled bargainers, 154, 156
Private corporations, 121
Problem census, 176
Problem solving, 176
Product name, 118
Products and technology risk information
        statement, 371–373
Profit, 102
Profit margin, 109
    great market squeeze, 193, 194f
Prospectus sample, 362–393
Provisional patents, 212–213
Public corporations, 121
Public market stage, 266
Public stock, 339–343
    sample prospectus, 362–393
Pull strategy, 181–182
Push strategy, 182

Q

Qualified market, 185
Quality, 191–192

R

Radical innovations, 104
Reduction to practice, 222
Referent power, 69
Regulatory risks information statement, 374–376
Representations and warranties, 348
Research & development
    intrapreneurship and academia, 37–38
    rules for Skunk Works, 18–20
Research analyst communications, 237–238
Resistance to change, 75, 76t
    overcoming inertia and defenses, 147–150
Resource poverty, 99–100
Return on investment (ROI), 257, 267
Revolutions, 93
Reward power, 69
Rewards, 84–85, 148f
Risk, 11, 57
Risk information statement, sample prospectus, 367–378
Risk managers, 57, 58f
Risk mitigation, 277
Royalties, 223

S

Sales, 181, 198
    complex, 193
    pricing, 190–193
    tensions with marketing, 192–193
Sales force compensation, 198–199
Sales force time management, 200–201
Sales promotions, 198
Sarbanes-Oxley Act, 228, 235
Schultz, Howard, 60
Science and technology gender gap, 42
Securities Act of 1933, 227, 234, 236–237, 363–364, 380–382, 384
Seed round, 257–258, 338
Selling security holders disclosure, 379–380
Serial entrepreneur, 86
Series "A" round, 135, 258–259, 338
Series "B" round, 135, 338
Series "C, D, E..." rounds, 339

Served market, 248
Service mark, 205
Silver, Spencer F., 21–23
Simply Green Biofuels, 47, 49
Situational Leadership Model (SLM), 63
Skills for entrepreneurs, 57, 58f
Skunk Works, Kelly's rules for, 18–20
Small Business Administration (SBA), 127
Small businesses, 2–3, *See also* Startup business ventures
    differences from big businesses, 99–100
    female-owned, 41, 47–50, *See also* Women and entrepreneurship
    loan approval rates, 45–46
Small Business Investment Act of 1958, 127
Smith, Frederick S., 60
Snow job, 169
Social media, 181
Soft bargainers, 154–155
Specialists, 93
Spin-offs, 30–31, 33–34
    equity carve-outs, 30, 34–36
Spin-outs, 30–33
    reasons for, 35t
Staged capital infusion, 266
Stages of organizational growth, 93–94
Stakeholders, creating value for, 123
Stalemate, 170
Standard of value, 262
Startup business ventures, *See also* Entrepreneurship; New venture creation; Small businesses
    board of advisors, 66–67, 137
    board of directors, 65–66, 137
    crises, 94
    estimated expenses, 133
    founding team, 64–65, 137–138
    growth models, 93–96
    importance of leadership, 61–62
    life cycle, *See* Corporate life cycles
    pioneer strategy, 190
    statistics regarding, 89
    strategic planning, 107–108
    success/failure, *See* Failure; Success
    trade secrets in, 225
    typical historical development, 128f

Stock dilution, 257–258
Stock offering ("going public"), 339–343
  sample prospectus, 362–393
Strategic alliances, 343–345, *See also*
      Mergers and acquisitions
  advantages, 345–346
  best practices, 347–349
  pitfalls, 347
Strategic management, 239
  competitive intelligence, 244–246
  core values, 242
  focus, differentiation, and cost
      leadership, 239
Strategic marketing goals, 250–252
Strategic planning, 107–108, 240–241
  SWOT analysis, 110*f*, 112–113
Strategy, 108, 239
Success
  defining, 43
  measuring in negotiations, 164
  merger and acquisition, 351–352
  successful entrepreneur characteristics,
      9*t*
  women entrepreneurs and, 43–45
Support activities, 109
Surgical techniques, 220
Sustainable business enterprises, 50
Sustaining innovations, 104
Sweat equity, 124–125, 256–257
SWOT analysis, 110*f*, 112–113, 357–361

T

Target market, 185, 248, 250
Target point in negotiations, 162
Teamwork, 281
Technology transfer, 37–38
Ten Commandments of, 38–39
ThermoFisher Scientific, 31–33
3M, 21–23
"Three envelopes" story, 150–151
Time, information, and reward (TIR)
      power, 173
Time management, 80–82, 170–171,
      200–201
Time Warner-AOL merger, 351–352

Total available market, 248
Trade dress, 205*t*, 206
Trademark, 205
Trade secrets, 205*t*, 206, 225
Transactional leadership, 63
Transformational leadership, 63
Treasurer, 122
Trump, Donald, 86

U

Unfreeze, change, refreeze model of
      change, 76–77
United States Patent Office (UPTO), 214
University-based technology transfer, 37–38
Urgent-important matrix, 81–82

V

Valley of Death, 114*f*, 115
Valuation, 134–136, 255, 260, *See also*
      Pricing
  asset-based or cost approach, 263
  capitalized returns method, 260*f*,
      261–263
  don't undersell, 52
  entrepreneur spin, 270
  examples, 135–136
  founder's syndrome of overvaluation,
      256
  income approach, 260–263
  market approach, 263
  maximal intrinsic value, 271
  negotiation and, 259
  Pre-Money and Post-Money, 134–135
  pre-seed financing, 256–257
  series "A" round, 258–259
  startup (seed round), 257–258
  stock dilution, 257–258
  value equation, 105*f*
  venture capital, 263–269
Value-based new products, 191–192
Value chain analysis, 108–109
Value proposition, 109–111
Values, 108, 242
Value vs. price, 256

Venture capital (VC), 46, 126–127, *See also* Financing
    female entrepreneurs and, 47
    intrapreneurship and academic research, 37–38
    investment criteria, 262*f*, 265
    investment hurdle rate, 268
    liquidity events, 264–265
    metrics, 267
    staged capital infusions, 266
    valuation, 263–269
    vulture capitalist, 269–270
    warning signs, 269
Vice president, 123
Vision, 107, 140, 145–146, 242–244
Visionaries, 188
Vulture capitalist, 269–270

W

Walk away price (WAP), 162
Wal-Mart, 3
Walton, Sam, 60
Wants, 184, 186

Warranties, 348
Welch, Jack, 59
"Why" questions of investors, 282
Win-lose negotiation, 156–157
Win-win negotiation, 157
Women and entrepreneurship, 41
    angel capital and, 47
    female-owned business example, 47–50
    global impact, 41
    interview with business owner, 51–54
    male and female entrepreneur characteristics, 42*t*
    motivations, 43
    respect and, 54
    science and technology gender gaps, 42
    small business loan approval rates, 45–46
    success and, 43–45
Worst alternative to a negotiated agreement (WATNA), 162

Z

Zone of possible agreement (ZOPA), 162

About the Author

Having conceptualized, financed, taken public, and run three public companies and founded and sold a fourth over the last 34 years, Dr. Michael Szycher is a successful entrepreneur who has created substantial value for his stakeholders over his career. As Chairman and/or CEO of these companies, he has been responsible for public offerings, acquisitions, product introductions, and spinouts. To strengthen his capabilities as an entrepreneur and business executive, he first obtained his PhD in Cardiac Physiology at Boston University School of Medicine, his Executive MBA from Suffolk University, and then supplemented that with a strong course of study in business law at Massachusetts School of Law.

Dr. Szycher is the editor of five books, *Biocompatible Polymers, Metals and Composites*; *Synthetic Biomedical Polymers*; *Blood Compatible Materials and Devices: Perspectives Towards the 21st Century*; *High Performance Biomaterials: A Comprehensive Guide to Medical/Pharmaceutical Applications*; *Szycher's Dictionary of Biomaterials and Medical Devices*, *Szycher's Dictionary of Medical Devices*, and *Szycher's Handbook of Polyurethanes*, considered the industry Bible at colleges and universities throughout the world.

Printed in the United States
by Baker & Taylor Publisher Services